The Sanskrit Astronomical Table Text *Brahmatulyasāraṇī*

Time, Astronomy, and Calendars

TEXTS AND STUDIES

VOLUME 9

The titles published in this series are listed at *brill.com/tac*

The Sanskrit Astronomical Table Text *Brahmatulyasāraṇī*

Numerical Tables in Textual Scholarship

By

Anuj Misra
Clemency Montelle
Kim Plofker

BRILL

LEIDEN | BOSTON

Cover illustration: Table of *manda*-equation (i.e. equation of centre) of the Sun in MS Smith Indic 45, fol. 6v. Reproduced with kind permission of the curators of the Smith Indic Collection, Columbia University, New York, USA.

Library of Congress Cataloging-in-Publication Data

Names: Misra, Anuj, author. | Montelle, Clemency, author. | Plofker, Kim, author.
Title: The Sanskrit astronomical table text Brahmatulyasāraṇī : numerical tables in textual scholarship / by Anuj Misra, Clemency Montelle, Kim Plofker.
Description: Leiden ; Boston : Brill, 2020. | Series: Time, astronomy, and calendars, 2211-632X ; volume 9 | Includes bibliographical references and index.
Identifiers: LCCN 2020036150 (print) | LCCN 2020036151 (ebook) | ISBN 9789004431416 (hardback) | ISBN 9789004432222 (ebook)
Subjects: LCSH: Hindu astronomy. | Astronomy, Ancient. | Astronomy–India–History. | Manuscripts, Sanskrit.
Classification: LCC QB18 .M57 2020 (print) | LCC QB18 (ebook) | DDC 520.954/09021–dc23
LC record available at https://lccn.loc.gov/2020036150
LC ebook record available at https://lccn.loc.gov/2020036151

Typeface for the Latin, Greek, and Cyrillic scripts: "Brill". See and download: brill.com/brill-typeface.

ISSN 2211-632X
ISBN 978-90-04-43141-6 (hardback)
ISBN 978-90-04-43222-2 (e-book)

This book is printed on acid-free paper and produced in a sustainable manner.

To our teachers who showed us the importance of astronomical tables,
to our colleagues tirelessly studying them, and
to our students carrying on the work.

किं वा गणका एवात्र प्रमादयुक्ता बभूवुः
किं वा लेखकपरंपरया कोष्टकेषु लिखितमशुद्धमस्ति । तद्गणकैः शोध्यते ॥

NITYĀNANDA, *Siddhāntasindhu*

Whether [because] mathematicians themselves in this case have become attached to error, or because of the succession of scribes, the writing in the tables is inaccurate. This is corrected by mathematicians.

Contents

CHAPTER 1

Introduction

The German literary critic Paul Maas summarised in his seminal *Textual Criticism* (Maas, 1956, p. 1) a traditional consensus about the primary aim of that discipline: 'The business of textual criticism is to produce a text as close as possible to the original (*constitutio textus*)'. In other words, an editor of a text generates, through careful and deliberate comparison of sources and their dependencies, an 'authoritative version' of the text which resembles the original composition as closely as possible. This endeavour implies the assumption that there once existed a unique and definite original text from which multiple generations of copies were subsequently created. In this process of transmission, errors and other alterations in the textual content were inadvertently (or sometimes purposefully) introduced. Consequently, the extant manuscript copies available to modern textual editors usually contain a complicated melange of variants. These discrepancies are employed to determine the stemmatic relationships (*stemma codicum*) among the copies. Criteria for inferring the nature and causes of textual variants are highly subjective and context-dependent, so different editorial reconstructions of an original text from the same set of manuscript witnesses may differ to a significant extent.[1] More recent scholarly perspectives, including ones frequently designated 'postmodern' or 'deconstructionist', have critiqued this traditional approach to textual editing. Some of them call into question the reality of the fundamental concepts of text and editorial, and even authorial, authority in the interpretation of texts. Other challenges that have perhaps had more practical impact on the work of editors have emerged from the rapidly expanding applications of digital technology to textual criticism, such as techniques of cladistic analysis, originally adapted from phylogenetics. The growth of so-called 'social editing' or 'wiki'-style collaborations in analysing and reading texts, likewise facilitated by digital technologies, has also influenced accepted ideas of what a textual edition should be.[2]

1 For this and other general studies on textual criticism see (Maas, 1956, Appendix I 42–49), (West, 1973), (Tanselle, 1996), (Van Hulle, n.d.); and (Trovato, 2014).

2 Several critiques of traditional scholarly editing are surveyed in (Greetham, 1993). Digital techniques for efficiently reconciling multiple variations and/or modifications within textual content and editing are comprehensively discussed in, e.g., (Burnard, O'Keeffe, and Unsworth, 2006) and (Driscoll and Pierazzo, 2016). For cladistic analysis, see, e.g., (Robins, 2007), and for social editing practices, (Robinson, 2016).

 | DOI:10.1163/9789004432222_002

All these developments have modified, though not rendered obsolete, the traditional aim of textual editing. An edited text nowadays may present simultaneously in digital form several related versions whose relative 'accuracy' or 'authority' the editor(s) do not presume to affirm definitively. But the traditional goal of reconstructing a linguistically coherent and historically plausible reconstruction of some presumed authorial intent still largely holds sway.

1.1 Critical Editing and Numerical Tables

The standard assumptions and strategies outlined above are challenged in a variety of ways by the task of creating critical editions of works largely composed of numerical data, such as astronomical tables.[3] For one thing, table texts associated with a certain title and/or author often exist in distinct recensions independently computed by different scribes or compilers, rather than being *verbatim et numeratim* copies descended from a unique original text. Even if their numerical data stay essentially the same from copy to copy, their row-and-column grids can be rearranged in many different ways to accommodate different desiderata such as size compression, user-friendliness, or topical coherence.

Other aspects of numerical tables which are subject to modification in different versions of the 'same' table text include table ordering within a multi-table set, surrounding 'paratext' (such as marginal notes, headings and titles, or accompanying textual excerpts), numerical precision and granularity, supplementary data such as numerical differences between successive table entries, the use of colour codes to identify different types of data, and the step size of the table argument. All these features may be altered at the hands of a new copier, possibly without substantial modification of the original mathematical algorithm used to produce the numerical data. The resulting issues concerning what constitutes a variant feature worth recording as opposed to a trivial discrepancy quickly become contentious. We present a summary of these issues in SEC. 4 and explain our own decisions concerning them in SEC. 5.

Variant readings in the numerical data itself also pose new problems in identifying and rationalising their different possible causes (van Dalen, 1993, pp. 12–18). Probably the most common source of discrepancies, endemic in all textual genres, is unintentional scribal (copying) errors. Computational 'errors',

3 For an overview of the general nature and history of astronomical tables, see (*Tables Analysis Method for the history of Astral Sciences* 2015). More detailed analyses of this genre in different scientific traditions are cited in section 1.2.

on the other hand, occur in data that have been mathematically recomputed or modified (not necessarily for the worse, as far as accuracy or consistency is concerned) rather than merely copied. Such modifications may involve changing the level of precision in numerical entries, adherence to different rounding conventions, incorporating a different epoch correction, alternative interpolation techniques for intermediate values, expanding or reducing the number of table entries, and so on. These types of changes complicate the fundamental problem of establishing an 'original' table from extant manuscript copies. An editor may describe, for example, two manuscript versions of the 'same' numerical table as variants of a common original, or else as two individual tables of somewhat different data.

In many cases, data from manuscripts can be compared with recomputed data, often revealing the table compiler's use of inexact (by modern standards) computation techniques that introduce what might be considered systematic errors. Unless the mathematical algorithm in question is very simple (such as repeated addition), modern computer algebra systems are unlikely to reproduce it correctly. Determination of trigonometric quantities, extraction of square roots, and the use of rounding or truncation at different points in the algorithm are some of the features of historical calculation that a typical computer algebra system routine fails to capture.

However, such features can be simulated by more carefully constructed computer programs that mimic hand computation practices. This sort of 'historically informed recomputation' uses decimal or sexagesimal integer digit arrays instead of native floating-point representations for numbers in arithmetic procedures. Likewise, transcendental function values such as sines are determined by interpolation in historically attested tables of the function, rather than by a complicated approximation algorithm. Exercising appropriate caution, data thus recomputed can be used to correct errors in the historical originals.

Ideally, such data reconstructions would include interactive versions where users could change the values of parameters, precision of values, rounding/truncating procedures, etc., in order to fine-tune the recomputing process. Statistical analysis should also be applied to both the recomputed and original data to identify artefacts of the methods used to generate them.

Related to, but separate from, the problem of reconstructing numerical table content is the reconstruction of algorithms and procedures for the use of the tables. As discussed below, table formats can be modified in different copies to facilitate different ways of using them, and accompanying verbal instructions are not always unambiguous about the steps to be followed. A complete manual of how to use a given table text is a desirable auxiliary to a critical edition of it, but cannot always be directly generated from the edition.

1.2 Textual Scholarship Applied to Table Texts: The State of the Field

Many of the issues described in the previous section have been addressed in various ways by editors of table texts in different linguistic and scientific traditions, from ancient Mesopotamia to the Neo-Latin manuscript corpus eventually supplanted by printed books. Modern digital humanities scholarship has initiated the creation of tools for analysing and editing tables, especially under the aegis of the recent 'Tables Analysis Method for the history of Astral Sciences' (TAMAS) project (Husson, 2015) and 'Digital Information System for the History of Astral Sciences' (DISHAS) project (Husson, 2018). In this section we briefly survey recent research and sketch some of the similarities and differences in approaches to editing numerical tables in different text corpora.

1.2.1 *Akkadian*

The earliest known astronomical tables in clearly identifiable row-and-column format, as well as the oldest archival documents containing such tables, occur in the Akkadian cuneiform corpus preserved on clay tablets from second-millennium BCE Mesopotamia. Much of the scholarship analysing them is surveyed in (Rochberg, 2016). The earliest of these are simple sequences recording, e.g., monthly variations in such quantities as the lengths of gnomon shadows or weights of water for a water-clock to measure a given fraction of the night. By the Late-Babylonian period prior to the turn of the Common Era, scientist-scribes were compiling very complex and extensive tables of computed function values; see, e.g., (Neugebauer, 1956, vol. I, p. 43). It might be assumed that since these are attested almost exclusively by the original contemporary documents, the conventional notion of critical editing—reconstructing and analysing the end results of a long process of scribal transmission—is largely irrelevant to scholars of this tradition. As noted in (Worthington, 2012, p. 3), however, this applies only to letters, lists and other ephemera on cuneiform tablets; Akkadian literary and scholarly texts were just as subject to the vagaries of repeated copying as those recorded in other languages on more perishable materials. Moreover, clay tablets' greater longevity does not confer immunity from physical damage such as chipping and breaking (and in the case of unbaked tablets, worm damage and salt incrustation as well). Some aspects of cuneiform editing and textual transmission of tables are discussed in (Neugebauer, 1956, vol. I, pp. 1–28) and (Ossendrijver, 2018).

1.2.2 *Greek*

Hellenistic Greek astronomy has left some contemporary documents in the form of fragmentary papyri (Jones, 1999), but the bulk of its corpus is attested only in later copies. Most of its surviving table texts other than the works of Ptolemy are small and incomplete. Even within this greatly reduced corpus, few items have yet been critically edited. The *Handy Tables* of Ptolemy (Tihon, 2011), (Mercier, 2011) are by far the most important extant exemplar in what remains of the Hellenistic or even the much larger Byzantine corpus of Greek astronomical table texts; most of the rest of this literature still awaits publication.[4]

1.2.3 *Chinese*

The application of critical editing techniques to Chinese texts has raised questions about the concept of 'original source' as a single authored text from a particular historical moment. Lancaster (Lancaster, n.d.) describes the editing process for ancient Buddhist works which

> were multiple streams of content (clusters of words) that flowed together, much like rivers that merge and create a single waterway and can then later diverge and split into delta like structures. The editing of Buddhist texts in such an environment is no longer just the search for the moment at which our document came into existence; it is rather a task of mapping the complexity of a series of text 'events'. This concept of how texts form is the opposite of the *Ur* approach which used the tree as a simile for textual development. The older approach was to assume that there was a single seed (autograph) from which the tree grew and the process of maturation produced a trunk which later branched into multiple limbs which in turn branched into stems and foliage. What the analytic software indicates for the Chinese Buddhist canon is that the simile must turn the tree upside down. The trunk is the extant example which we have before us in the form of a manifestation of the text. The trunk has occurred because of the many limbs and stems of past input.

The textual traditions of Chinese canonical astronomical systems (*li*) with their tables, on the other hand, may perhaps be better represented by stems of bamboo. Solidly codified as official state documents from the time of their construction, they were carefully transmitted as authoritative texts subject to little change, especially after the supersession of manuscript copying by printing

4 There are also editions of the late medieval Greek translations of Islamic *zījes* by Gregory Chioniades (Pingree, 2005), and a roughly contemporary Greek almanac that draws from the same sources (Mercier, 1994). See (Bardi, 2018) for an overview of recent scholarship on the Byzantine corpus.

around the start of the second millennium. Modern editions of the standard histories containing these *li* discuss the history of their textual scholarship; *vide*, for instance, (Cullen, 2011, 2016), (Chemla, 2016a,b,c), (Li, 2014, 2016), and (Morgan, 2016).

1.2.4 *Arabic/Persian*

The process of textual scholarship in Arabic science has more closely paralleled the philological tradition for classical works, largely because of the genetic connections between Greek and Arabic scientific texts (Carter, 1995). The study of *zījes* and other astronomical table texts in Arabic or Persian has gravitated toward broad surveys focused on their identification and interrelationships (Kennedy, 1956), (van Dalen, 1993), (King and Samsó, 2001), (King, 2004, vol. 1). Their general characteristics and their transmission within and beyond the Islamic world are discussed in (Van Brummelen, 2014); however, critical editions of *zījes* as individual works, beginning with (Nallino, 1899–1907), are accumulating only slowly (Samsó, 2003). Islamic *zījes* are not only one of the most prolific but arguably one of the most technically complicated and elaborate genres of astronomical tables, which makes undertaking such editions a very laborious task. The large-scale 'Ptolemaeus Arabus et Latinus' (PAL) project (Hasse, Juste, and van Dalen, 2013) has embarked on this venture for Arabic versions of Ptolemy's own works, but there are hundreds of other Islamic astronomical tables that need the same treatment.

1.2.5 *Latin*

The Arabic table text genre in its turn helped shape its Latin counterpart, as illustrated by (Bjørnbo, Besthorn, and Suter, 1914) and (Pedersen, 2002). The Ptolemaic corpus in Latin, including Latin versions of the *Handy Tables*, is also in the process of analysis and publication by PAL. The works of late medieval and early modern Alfonsine astronomy are the subject of the recently launched 'Alfonsine Astronomy' (ALFA) project (Husson, 2017). Latin astronomical tables in general throughout this period are surveyed in (Chabás and Goldstein, 2012), but most of the extant manuscripts still await individual critical editions.

1.2.6 *Sanskrit*

Much still remains to be learned about the history of the genre of sets of astronomical tables (*koṣṭhaka, sāranī*) in Sanskrit. Some seem to have served (or at least originated) as supplementary material to the versified instructions and data contained in an existing handbook or treatise. Others appear to have been conceived and created as standalone texts with independently designed formats and procedures. Early printed versions of some *koṣṭhakas*, such as

(Makaranda, 1923) and (Pandeya, 1938), attest to both their historical significance and their continuing importance to practitioners of *jyotiṣa* or Sanskrit astral sciences.

Large-scale historical investigation of this genre, mostly still confined to manuscript form, goes back to Pingree's groundbreaking surveys and classifications in (Pingree, 1968) and (Pingree, 1973), and continues in, e.g., (Pingree, 1970–1994) and (Montelle and Plofker, 2018). The detailed analysis of individual table texts initiated in (Neugebauer and Pingree, 1967) has likewise been pursued further in several studies such as (Sarma, 2013), (Rupa K., Venugopal, and Rao, 2014) and (Montelle, 2014). Attempts to subject them to the process of critical textual editing, as in (Misra, Montelle, and Plofker, 2016) and the present work, must confront the challenges discussed in the following section as well as those peculiar to the nature of table texts.

1.3 Critical Editing and the Sanskrit Text Corpus

Critical editing of Sanskrit texts in all genres is in many ways more problematic than its counterpart in other text corpora. For one thing, the surviving manuscripts of Sanskrit works are far more numerous than Greek and Latin ones. They also contain a higher proportion of metrical verse to prose, and therefore more opportunities to test a surviving text against the guidelines of metrical regularity. Sanskrit palaeography at present is still a very understudied field, with many open questions about fundamental characteristics of various scripts and their temporal and regional ranges, the evolution of particular scribal conventions, etc. The prosopography of scribes and other copiers of surviving Sanskrit manuscripts is even more obscure. These matters are further complicated by the fact that Indic manuscript formats tend to be less standardised and strictly regulated than those of, e.g., Islamic or medieval Latin manuscripts.

Katre (Katre, 1954, p. viii) notes that 'the science of textual criticism as developed by Europeans does not solve all our Indian problems'. Not only the abovementioned abundance and uncertainties pertaining to Sanskrit texts, but also the diversity of writing media from birch-bark or palm-leaf to copper plates and paper, as well as the interposition of oral recitation between original composition and subsequent copying, pose challenges that the typical editor of western classical texts has not encountered.[5]

5 Sanskrit critical editing is discussed most fully so far in (Katre, 1954); see also (Rocher, 1995), (Wujastyk, 1993, vol. 1, xvi–xvii; vol. 2, xxix), (Slaje, 2008), (Pecchia, forthcoming).

At present, these challenges have been most fully explored (though by no means fully overcome) in the projects for creating editions of the great Sanskrit epics and other sacred and literary works. The editor of almost any *śāstra* or learned didactic text, including mathematical and astronomical ones, faces a task far less immense but still daunting. Although the numbers of manuscripts and recensions of such a work are usually nowhere near as overwhelming as in the case of the great epics, they can still be strongly divergent in their content and language, as well as confusing in their relationships.

CHAPTER 2

Overview of the *Brahmatulyasāraṇī* and Its Manuscripts

2.1 The *Brahmatulyasāraṇī*: Background and Approach

In this study, we explore some of the abovementioned issues through an anonymous Sanskrit astronomical table text from the mid-second millennium CE called the *Brahmatulyasāraṇī*, literally 'Tables of/for the *Brahmatulya*'. The tables and their accompanying versified instructions are based primarily on parameters and procedures from the astronomical handbook *Karaṇakutūhala* (*KKu*), also known as *Brahmatulya*, composed by Bhāskara II (Bhāskarācārya) in the late twelfth century. The *Karaṇakutūhala* in its turn condenses and approximates many of the computational formulae in Bhāskara's earlier treatise, the magisterial *Siddhāntaśiromaṇi* (*SiŚi*) following the Brāhmapakṣa sunrise-epoch astronomical school. Bhāskara adapts these formulae for the *Karaṇakutūhala*'s epoch of sunrise on Thursday 1 Caitra Śaka 1105 (equivalent to 24 February 1183 in the Julian calendar, although an alternate reconstruction gives Wednesday 23 February of that year). The *Brahmatulyasāraṇī*'s own date of composition, as opposed to the 1183 epoch date of its source, is unknown; the incomplete available information about the dates of its known surviving manuscripts suggests that the tables were originally compiled in the sixteenth or seventeenth century.[1]

1 For a survey of the *Brahmatulyasāraṇī* and its manuscripts, including ones we have not consulted, see (Pingree, 1968, pp. 36–37). Its instructional verses were edited and analysed in (Montelle and Plofker, 2015), which also contains a fuller discussion of its date(s); a revised and updated version of this edited text appears in section 6.1 of the present work. Its relation to the *Karaṇakutūhala* is analysed in (Montelle and Plofker, 2018, pp. 201–211).

For the *Karaṇakutūhala* itself, see the bibliographic information in (Pingree, 1970–1994, A4, 322), the edited text in (Mishra, 1991), and the English translation and study in (Rao and Uma, 2008). The issue of the *Karaṇakutūhala*'s epoch is addressed in (Pingree, 1970–1994, A4, 322), (Sarma, 2019, p. 23), (Rao and Uma, 2008, p. iv), and (Plofker, n.d.) on *Karaṇakutūhala* 1.2–3.

For the *Siddhāntaśiromaṇi*, see (Pingree, 1970–1994, A4, 311) and (Śāstrī, 1989); many detailed derivations and expositions of its astronomy and mathematics may be found in studies such as (Arkasomayaji, 1980) and (Hayashi, Montelle, and Ramasubramanian, 2019). Bhāskara's adaptation of *Siddhāntaśiromaṇi* formulae to *Karaṇakutūhala* approximations is also discussed in, e.g., (Plofker, 2016).

 | DOI:10.1163/9789004432222_003

The *Brahmatulyasāraṇī* draws on the first two chapters of the *Karaṇakutūhala* concerning the mean motions of the planets, with the appropriate geographic corrections and corresponding orbital anomaly corrections for determining their true motions and positions. It is not entirely clear whether the unknown architect of the *Brahmatulyasāraṇī* intended it as a supplement to or a substitute for the *Karaṇakutūhala*. Most of its procedures for these planetary computations closely follow those of the *Karaṇakutūhala*, but it includes some notable deviations discussed in chapter 3, as well as its own versified set of instructions.

The remainder of this book consists of our critical edition and study of the *Brahmatulyasāraṇī*, divided into the following parts:

- Sections 2.2 and 2.3, describing the manuscripts we used and discussing the information in their colophons and post-colophons;
- Chapter 3 containing a summary of all the *Brahmatulyasāraṇī* tables and an analysis of their content and the procedures for their use. This chapter, following the conventions described in Appendix, section 1, includes transliterations, translations and identifications of the metrical forms of the work's instructional verses. Descriptions of the verses' corresponding tables attempt to reconstruct how the former would have guided users through computations involving the latter;
- Chapter 4 on the details of the individual tables and the different ways they, and their accompanying paratext such as headings and notes, are written and arranged in the different manuscripts of the *Brahmatulyasāraṇī*;
- Chapter 5 describing our editorial conventions and the issues they are intended to address in presenting the critical edition;
- Chapter 6, the critical edition of the versified text and tables of the *Brahmatulyasāraṇī*;
- Appendix, containing supplementary information about text representation, numerical notation and calendar reckoning in Sanskrit astronomy, as well as analyses of the *Karaṇakutūhala* algorithms underlying the *Brahmatulyasāraṇī* tabular data.

2.2 Manuscript Witnesses to the *Brahmatulyasāraṇī*

In preparing our edition we relied on photocopies and digital images of the manuscripts listed in the following table. These reproductions were obtained from the John Hay Library at Brown University, USA, the Rare Book and Manuscript Library at Columbia University, USA, and the Bhandarkar Oriental Research Institute in Pune, India. All manuscripts are written in *nāgarī* script on hand-made paper, with table heading and row header text in Sanskrit. The tables themselves are oriented in a landscape format, with the table argument running horizontally and tabulated function values written vertically. Successive tables or continuations of the same table are stacked vertically on the page. Some of the accompanying paratext is in an unidentified vernacular that we have not been able to transliterate or translate with confidence and consistency. Due to the dearth of (what we are able to identify as) consistent or recognisable palæographic traits in the mostly numerical manuscripts, we have not attempted to construct a manuscript stemma for them.

Siglum	Shelfmark	Folios	Contents
B	BORI 501/1895–1902	ff. 2–28	Tables
Kh	Khasmohor 5424 (a,b)	(a): f. 1	Tables and text
		(b): ff. 1–13	Tables
S29	Poleman 4952 (Smith Indic 29)	f. 6	Text
S43	Poleman 4876 (Smith Indic 43)	ff. 1–13	Tables
S45	Poleman 4735 (Smith Indic 45)	ff. 2–17	Tables
SMB	Poleman 4946 (Smith Indic MB LVIII)	ff. 1–5	Tables

A brief description of each of the individual manuscripts follows, including a citation of the manuscript catalogue containing its available codicological details.

MS B BORI 501/1895–1902, ff. 2–28. (Bhandarkar Oriental Research Institute, 1990–1991, pp. 70–71)

Begins on f. 2r (recto) with Table I (*vide* Schema 1). In the black-and-white photocopies available to us, the manuscript appears legibly written with occasional ink blots and smudges, between double-lined margins, *vide* figure 1. Table grids sometimes include blank cells or extra space, and table headings sometimes spill into the margins. The verso of each folio has the page number written in the bottom right corner (in the outer margin) of the folio. The date Saṃvat 1734, Kārttika *śuklapakṣa* 2, Wednesday (≈ Thursday 28 October 1677? assuming expired years) appears on f. 28v (verso).

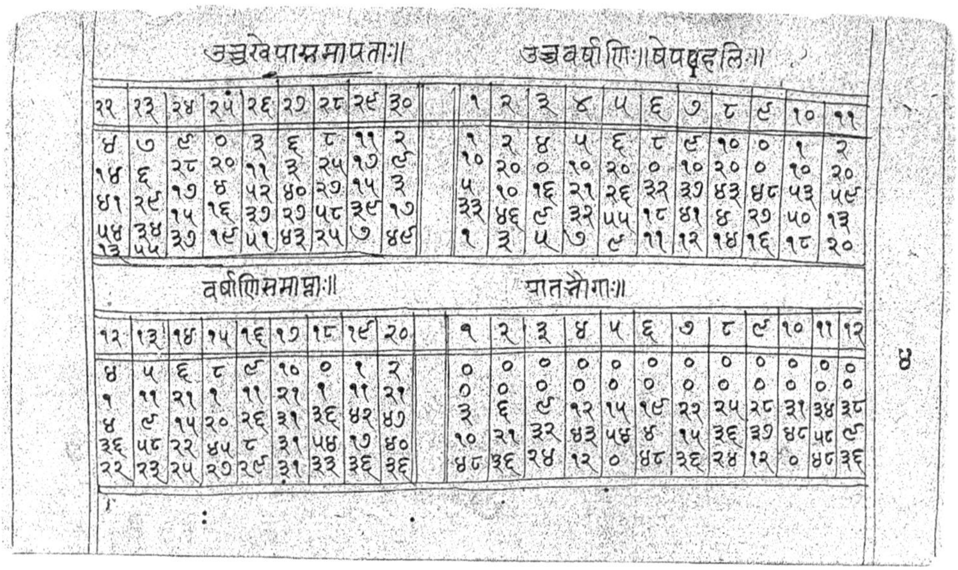

FIGURE 1 A sample image of MS B f. 4[v] showing the general layout of the manuscript

MS Kh Khasmohor 5424 (a,b), f. 1rv + ff. 1–13. (Pingree, 2003, pp. 47–48, 49–51)

The copy of MS Kh available to us includes f. 1[rv] of Khasmohor 5424(a) (henceforth Kh (a)) containing tables and text, and ff. 1–13 of Khasmohor 5424(b) (Kh (b)) containing tables. In the black-and-white photocopies available to us, the manuscript is legibly written with no ink blots or smears. The paratext alongside the tables on f. 1[v] of Kh (a) is written in an unidentified vernacular language. The tables on ff. 1–13 of Kh (b) are written between (faint) double-lined margins, *vide* figure 2. Occasionally, the table headings and table columns extend into the margins of the folio. The verso of each folio has the page number written in the bottom right corner (in the outer margin) of the folio; however, the sequence of numbering is irregular. On f. 1[v] of Kh (a) we find the word *śivaḥ* written above the folio number 1 in the right margin. On the same folio, the paratextual material associated with the tables appears to be written in an unidentified vernacular language. Folio 13[v] of Kh (b) has an incomplete table of *traikya* for 1°–107° (not included in our edition; *vide* Schema 1), below which is written '*atha rājamṛgāṅkoktāni traikyāni*', referring to the *Rājamṛgāṅka* of Bhojadeva (1042 CE) (Pingree, 1987). Paratextual material appears in the left and right margins of this folio, and at the top of it '*jyotisasāraṇīpatra* 13' is written in a different hand.

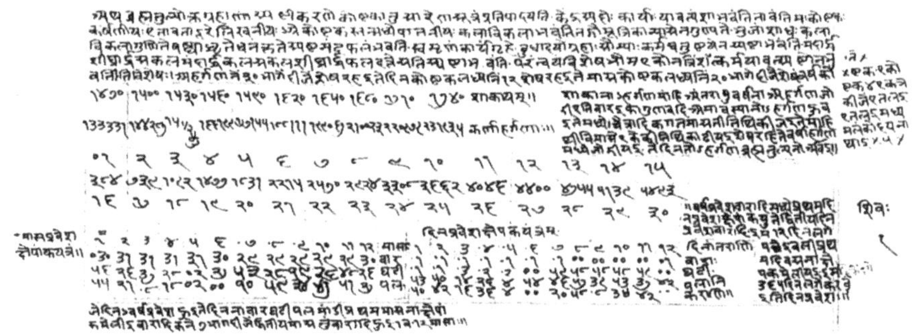

FIGURE 2 A sample image of MS Kh f. 1v showing the general layout of the manuscript

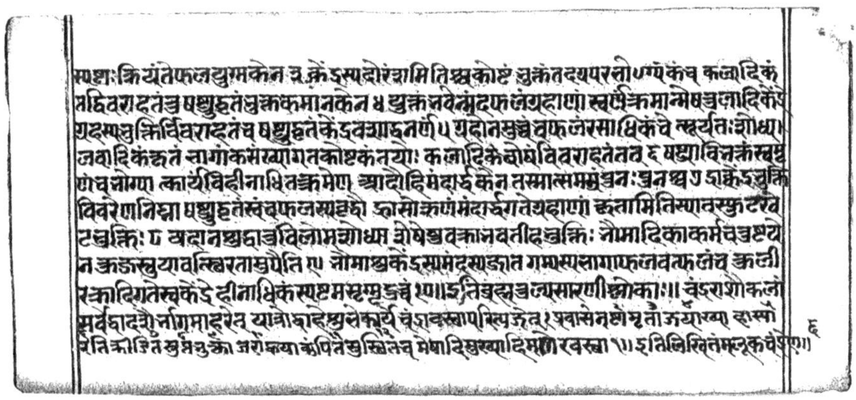

FIGURE 3 A sample image of MS S29 f. 6v showing the layout of the verse table-text of the *Brahmatulyasāraṇī*

MS S29 Poleman 4952 (Smith Indic 29), ff. 1–6. (Pingree, 2007, p. 78)

The last folio (f. 6rv) contains the *Brahmatulyasāraṇī*'s short verse text legibly written between double-lined margins; *vide* figure 3. The first five folia (ff. 1–5) contain numerical tables derived not from the *Brahmatulyasāraṇī* but from a different work, the *Candrārkī* of Dinakara, whose epoch is Śaka 1500 (1578 CE) (Kolachana et al., 2018). Folio 6v also contains the colophon and post-colophon, including an astrological verse and the identification of the (otherwise unknown) scribe as Malūkacandra. See SEC. 2.3 and (Pingree, 1968, p. 34).

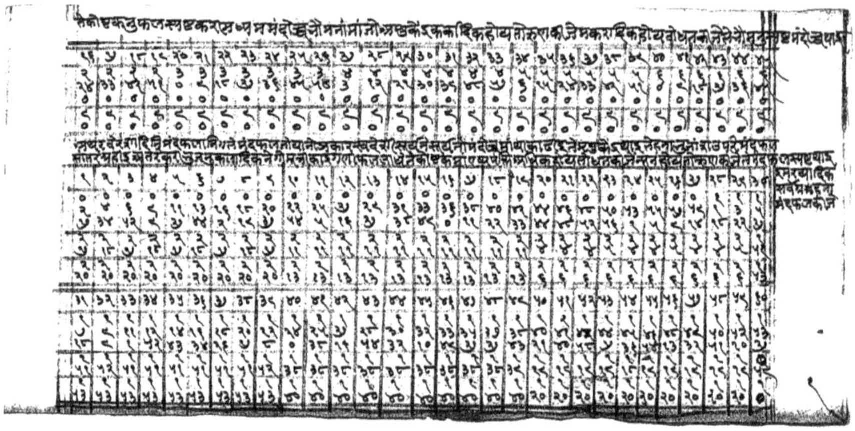

FIGURE 4 A sample image of MS S43 f. 1ᵛ showing the general layout of the manuscript

MS S43 Poleman 4876 (Smith Indic 43), ff. 1–13. (Pingree, 2007, p. 58)

Begins on f. 1ʳ with Table X (*vide* Schema 1). Black-and-white photocopies with a few ink blots and smears. The tables are written between double-lined margins with table headings occasionally spilling into the outer margins, *vide* figure 4. Similar to MS B, the table grids sometimes include blank columns and the page numbers are written on the folio verso in the bottom right corner. The paratext accompanying some of the tables is written in an unidentified vernacular language. Folio 13ᵛ has a note in a different handwriting on the top left, adjacent to the table heading, that reads *śloka* 300 'verse 300'.

MS S45 Poleman 4735 (Smith Indic 45), ff. 2–17. (Pingree, 2007, pp. 57–58)

Begins on f. 2ʳ with Table I (*vide* Schema 1). Black-and-white photocopies showing features similar to MS S43, *viz.* clear legibility with occasional ink-smears, inter-tabular empty columns, and bottom-right marginal page numbers on verso. The tables themselves are written between triple-lined margins with paratext (in Sanskrit or an unidentified vernacular language) along the outer margins, *vide* figure 5. Folio 1ʳ has a note in a different handwriting on the top right, adjacent to the table heading, that reads *śloka* 600 'verse 600'. The date Monday Kārttika *kṛṣṇapakṣa* 11, current [year] Saṃvat 1855/ Śaka 1720 (= Monday 16 October 1797?, assuming current years and *pūrṇimānta* month) appears on f. 17ᵛ.

FIGURE 5 A sample image of MS S45 f. 6^{v} showing the general layout of the manuscript

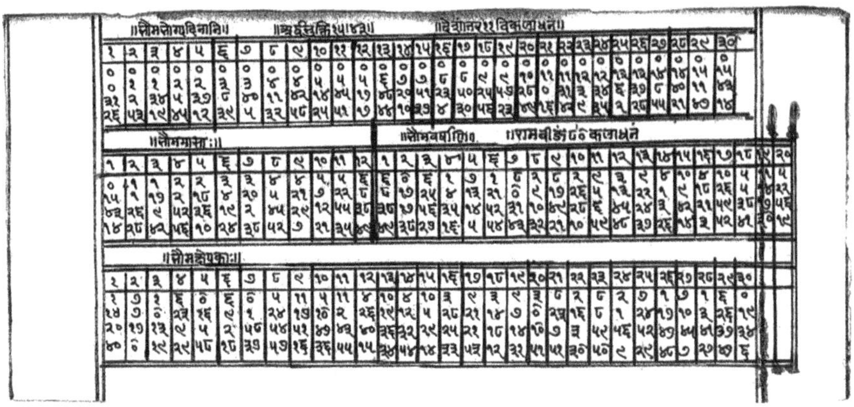

FIGURE 6 A sample image of MS SMB f. 5^{r} showing the general layout of the manuscript

MS SMB Poleman 4946 (Smith Indic MB LVIII), ff. 1–5. (Pingree, 2007, pp. 58–59)

Begins on f. 1^{r} with Table I (*vide* Schema 1). Black-and-white photocopies showing features similar to MS S43, *viz.* clear legibility with occasional ink-smears, inter-tabular empty columns, and bottom-right marginal page numbers on verso. The tables themselves are written between triple-lined margins with paratext (in Sanskrit or an unidentified vernacular language) along the outer margins, *vide* figure 6. Folio 1^{r} has a note in a different handwriting on the top right, adjacent to the table heading, that reads *śloka* 160 'verse 160'.[2]

2 MS SMB_2 = Poleman 4946 (Smith Indic MB LXII) is a single folio (f. 11rv). In (Pingree, 1968,

2.3 Colophon and Post-colophon Material from the Manuscripts

Manuscripts B, Kh(a), S29 and S45 contain colophon and/or post-colophon text. Since this textual material does not form part of our critical edition, we include here a transcription of its *nāgarī* form as well as its transliteration and translation.

MS B (f. 28v)

इति श्रीकर्णकुतूहले मन्दशीघ्रफलं संपूर्णं समाप्तामिति ॥ शुभं अस्तुः ॥
संवत् १७३४ वर्षे काती सुदी २ बुधवारे पोथी लषीतं चरंबगसु ॥

iti śrīkarṇakutūhale mandaśīghraphalaṃ sampūrṇaṃ samāptām iti ‖ *śubhaṃ astuḥ* ‖ *saṃvat 1734 varṣe kātī sudī 2 budhavāre pothī laṣītaṃ caraṃbagasu* ‖

Thus, [the description of] the *manda*- and *śīghra*-equations in *Śrī Karaṇakutūhala* is fully completed. May it be auspicious. In the year Saṃvat 1734, Kārttika *śuklapakṣa* 2, Wednesday [≈ Thursday 28 October 1677 assuming expired years], the book was written (*laṣītaṃ* for *likhitaṃ*?) in Charam Bag [?].

MS Kh(a) (f. 1r)

इति श्रीकर्णकुतूहले विदग्धबुद्धिवल्लभस्पष्टाधिकारम् ॥ श्रीरघुनाथाय नमो नमः ॥

iti śrīkarṇakutūhale vidagdhabuddhivallabhaspaṣṭādhikāram ‖ *śrīraghunāthāya namo namaḥ* ‖

Thus, the chapter on true positions dear to [those] of sharp intellect. Repeated salutations to Lord Raghunātha.

MS S29 (f. 6v)

इति ब्रह्मतुल्यसारणीश्लोकाः ॥ *iti brahmatulyasāraṇīślokāḥ* ‖

Thus, the *ślokas* of the *Brahmatulyasāraṇī*.

p. 36), MS SMB_2 is mistakenly identified as Poleman 4946 (Smith Indic MB) LXII f. 11, while on p. 32 of (Pingree, 1968), it is stated to be f. 11rv of MS S45.

चन्द्रराशौ कलां सर्वे	*candrarāśau kalāṃ sarve*
द्वादशैर्भागमाहरेत्	*dvādaśair bhāgam āharet*
यात्रोद्वाहे शुभे कार्ये	*yātrodvāhe śubhe kārye*
चन्द्रावस्थाः परित्यजेत् १	*candrāvasthāḥ parityajet* 1
१ २ ३ ४	*1 2 3 4*
प्रवासनष्टामृताजयाख्या	*pravāsanaṣṭāmṛtatājayākhyā*
५ ६ ७ ८ ९	*5 6 7 8 9*
हास्यारतिक्रीडितसुप्तभुक्ता ।	*hāsyāratikrīḍitasuptabhuktāḥ*
१० ११ १२	*10 11 12*
जराह्वया कंपितसुस्थितं च	*jarāhvayāḥ kampitasusthitaṃ ca*
मेषादिमुख्या हिमगोरवस्थाः १ ॥	*meṣādimukhyā himagor avasthāḥ* 1 ‖

In every zodiacal sign of the Moon, one should divide the minute [and?] degree [? of longitude] by twelve. [The remainder gives the number of the lunar *avasthā* or 'status.'] When a journey or marriage is to be made auspicious, the *avasthā* of the Moon should be disregarded. [Those] called journey, loss, immortality, victory; laughter, delight, play, sleep, eating; [that] called old age, and trembling [and] comfort; are the states (*avasthās*) of the Moon starting from the beginning of Aries.

इति लिखितं मलूकचन्द्रेण ॥	*iti likhitaṃ malūkacandreṇa* ‖

Thus, [this text] was written by Malūkacandra.

MS S45 (f. 17^{v})

इति श्रीब्रह्मतुल्यस्य ग्रहसाधनार्थं सारणी सम॰ ॥ सारणीसमाप्त ॥

संवत् १८५५ वर्षे शाके १७२० प्रवर्त्तमने कार्त्तिकविद ११ सोमे ॥

iti śrībrahmatulyasya grahasādhanārthaṃ sāraṇī sam° ‖
sāraṇīsamāpta ‖
saṃvat 1855 varṣe śāke 1720 pravarttamane kārttikavida[3] *11 some* ‖

3 We read *kārttikavida* as a scribal error for *kārttikavadi* meaning the *kṛṣṇapakṣa* 'dark fortnight' of the month of Kārttika.

Thus, the tables for the sake of explaining the computations of the planets belonging to *Śrī Brahmatulya* are com[plete]. End of the tables. [Written] in the current year Saṃvat 1855, Śaka 1720, Kārttika *kṛṣṇapakṣa* 11, Monday [= Monday 16 October 1797, assuming current years and *pūrṇimānta* month].

CHAPTER 3

Technical Analysis of the *Brahmatulyasāraṇī*

3.1 Overview of the *Brahmatulyasāraṇī* and Its Tables

The intended purpose of the *Brahmatulyasāraṇī* is described in its introductory verse:

> *oṃ* ‖ *śrīgaṇeśāya namaḥ* ‖
>
> *natvā vallabhanandanaṃ tadanugopālāṃhripadmadvayaṃ*
> *jñātvā śrīguruvākyato hy aharniśaṃ sadyuktim evādhunā* ‖
> *siddhānteṣu yathoktakhecaravidhiḥ suspaṣṭakoṣṭhaṃ muhur*
> *madhyaspaṣṭavibhāgato grahagaṇāt kurve dinaughād aham* ‖ 1 ‖
> *śārdūlavikrīḍita*
>
> VERSE 1
>
> OṂ. Homage to Lord Gaṇeśa.
>
> Saluting Vallabhanandana and after him the two lotus feet of Gopāla,[1] having learned the true method from the word of the revered teacher by day and night; now, the rule of the planets, as spoken in the *siddhāntas*. I shall compute a very accurate set of tables separately for mean and true [quantities] for the various planets, from the accumulated days.

Here the unidentified author of the *Brahmatulyasāraṇī*, after a pious invocation, declares his intention to provide tables for planetary positions in accordance with the authority of the *siddhāntas* or canonical treatises of mathematical astronomy.[2] The content and organisation of these tables, based on

1 The invocation of Vallabhanandana, which baffled us in our initial study (Montelle and Plofker, 2015, v. 1 on pp. 6–7), may refer to Viṭṭhalanātha Gosvāmī (1516–1586), the son of the founder Vallabhācārya of the Vaiṣṇavite Puṣṭimārga tradition. If correct, this identification would fit in well with the accompanying invocation of Kṛṣṇa-Gopāla, especially revered in Puṣṭimārga. *Vide* (Smith, 2011, p. 179) and (Saha, 2004, pp. 118–128).

2 The emergence and evolution of authoritative textual traditions in Sanskrit astronomy has a long and complex history, many facets of which are discussed in, e.g., (Ruggles, 2015, Vol. 3, Part XII).

 | DOI:10.1163/9789004432222_004

their witnesses in MSS B, Kh, S43, S45, and SMB, are outlined in Schema 1. The remainder of this chapter explains what we know and conjecture about their construction and use, derived from the content of the tables themselves, the accompanying verse instructions, and the portions of the *Karaṇakutūhala* (*KKu*) and *Siddhāntaśiromaṇi* (*SiŚi*) that originally inspired them. For detailed expositions of the mathematical-astronomy formulae upon which the tables are based, *vide* Appendix, section 2.

The following discussion is laid out according to what we consider to be a logical sequence of steps for computing planetary positions for a desired time and locality by means of the tabulated data. The reader should bear in mind that this sequence is one of many possible variant interpretations rather than a self-evident reconstruction. As Schema 1 reveals, the available manuscripts of the *Brahmatulyasāraṇī* differ significantly in their choices about the number, content and ordering of the tables. The extremely concise verse instructions skim over many of the details of their use, as well as differing in some places from the corresponding algorithms in the *Karaṇakutūhala*. Moreover, we do not have any textual record, in the form of illustrative worked examples, of how its historical users employed these tables to perform specific computations. Hence, we have not attempted here to invent such examples with arbitrarily selected historical data.[3]

SCHEMA 1 Distribution of tables across folia of each of the manuscripts

Number	Type of table	Manuscripts				
		B	Kh	S43	S45	SMB
I	Mean longitudinal displacement of the Sun	ff. 2^r–2^v	f. 1^r	–	f. 2^r	f. 1^r
II	Mean longitudinal displacement of the Moon	ff. 2^v–3^v	f. 1^v	–	f. 2^v	f. 1^v
III	Mean longitudinal displacement of the Moon's apogee	ff. 3^v–4^v	ff. 1^v–2^r	–	f. 3^r	f. 2^r
IV	Mean longitudinal displacement of the Moon's node	ff. 4^v–5^v	ff. 2^r–2^v	–	f. 3^v	f. 2^v
V	Mean longitudinal displacement of Mars	ff. 5^v–6^v	ff. 2^v–3^r	–	f. 4^r	f. 3^r
VI	Mean longitudinal displacement of Mercury's *śīghra*	ff. 6^v–8^r	ff. 3^r–3^v	–	f. 4^v	f. 3^v
VII	Mean longitudinal displacement of Jupiter	ff. 8^r–9^r	ff. 3^v–4^r	–	f. 5^r	f. 4^r
VIII	Mean longitudinal displacement of Venus' *śīghra*	ff. 9^r–10^r	ff. 4^r–4^v	–	f. 5^v	f. 4^v

3 Note that genuine versions of illustrative worked examples do exist for the *Karaṇakutūhala*, e.g., in the commentary of Sumatiharṣa edited in (Mishra, 1991) and frequently cited in (Rao and Uma, 2008).

SCHEMA 1 Distribution of tables across folia of each of the manuscripts (*cont.*)

Number	Type of table	Manuscripts				
		B	Kh	S43	S45	SMB
IX	Mean longitudinal displacement of Saturn	ff. 10^{r}–11^{r}	ff. 4^{v}–5^{r}	–	f. 6^{r}	f. 5^{r}
X	*Manda*-equation of the Sun	ff. 11^{r}–12^{r}	ff. 5^{r}–5^{v}	ff. 1^{v}–2^{r}	f. 6^{v}	–
XI	*Manda*-equation of the Moon	ff. 12^{r}–13^{r}	ff. 5^{v}–6^{r}	ff. 2^{r}–2^{v}	f. 7^{r}	–
XII	*Manda*-equation of Mars	ff. 13^{r}–14^{v}	ff. 6^{r}–6^{v}	ff. 2^{v}–3^{r}	f. 8^{r}	–
XIII	*Manda*-equation of Mercury	ff. 16^{v}–17^{v}	ff. 6^{v}–7^{r}	ff. 3^{r}–3^{v}	f. 10^{r}	–
XIV	*Manda*-equation of Jupiter	ff. 19^{v}–20^{v}	ff. 7^{r}–7^{v}	ff. 4^{r}–4^{v}	f. 12^{r}	–
XV	*Manda*-equation of Venus	ff. 22^{v}–23^{v}	f. 7^{v}	ff. 4^{v}–5^{r}	f. 14^{r}	–
XVI	*Manda*-equation of Saturn	ff. 26^{r}–26^{v}	f. 8^{r}	ff. 5^{r}–5^{v}	f. 16^{r}	–
XVII-A	*Śīghra*-equation of Mars	ff. 14^{v}–16^{v}	–	–	–	–
XVII-B		–	ff. 8^{v}–9^{v}	–	–	–
XVII-C		–	–	ff. 6^{r}–7^{v}	ff. 8^{v}–9^{v}	–
XVIII-A	*Śīghra*-equation of Mercury	ff. 17^{v}–19^{v}	–	–	–	–
XVIII-B		–	ff. 9^{v}–10^{v}	–	–	–
XVIII-C		–	–	ff. 8^{r}–9^{r}	ff. 10^{v}–11^{v}	–
XIX-A	*Śīghra*-equation of Jupiter	ff. 20^{v}–22^{v}	–	–	–	–
XIX-B		–	ff. 10^{v}–11^{r}	–	–	–
XIX-C		–	–	ff. 9^{v}–10^{v}	ff. 12^{v}–13^{v}	–
XX-A	*Śīghra*-equation of Venus	ff. 23^{v}–25^{v}	–	–	–	–
XX-B		–	ff. 11^{v}–12^{r}	–	–	–
XX-C		–	–	ff. 11^{r}–12^{r}	–	–
XX-D		–	–	–	ff. 14^{v}–15^{v}	–
XXI-A	*Śīghra*-equation of Saturn	ff. 27^{r}–28^{v}	–	–	–	–
XXI-B		–	ff. 12^{v}–13^{r}	–	–	–
XXI-C		–	–	ff. 12^{v}–13^{v}	ff. 16^{v}–17^{v}	–
XXII	Correction for Mars' *manda*-apogee	–	ff. 8^{r}–8^{v}	ff. 1^{r}–1^{v}	f. 7^{v}	
XXIII-A	'Rising-difference' (*udayāntara*) correction for	–	–	f. 1^{r}	–	–
XXIII-B		–	–	–	f. 7^{v}	–
XXIII-C		–	–	–	–	f. 5^{v}
XXIV-A	'Rising-difference' (*udayāntara*) correction for	–	–	f. 1^{r}	–	–
XXIV-B	the Moon	–	–	–	f. 7^{v}	–
XXIV-C		–	–	–	–	f. 5^{v}
XXV	Table of accumulated civil days (*ahargaṇa*)	–	f. 1^{v} MS Kh (a)	–	–	–
XXVI	Beginnings of solar months and solar days	–	f. 1^{v} MS Kh (a)	–	–	–
XXVII	Epoch positions of planetary apogees, longitudinal corrections (*deśāntara*) to mean planetary positions, and adjustments (*bīja*) to mean planetary positions	–	f. 8^{v}	–	–	–
XXVIII	Solar declination	–	ff. 13^{r}–13^{v}	–	–	–

3.2 Accumulated Civil Days (*ahargaṇa*) since Epoch; Mean Longitudes

3.2.1 *Computation of the* ahargaṇa *or Time since Epoch*

As indicated at the end of verse 1 of the text instructions above, the first step in any calculation involving mean longitudes in Sanskrit astronomy is the computation of the so-called *ahargaṇa*, or accumulated civil days from the text's epoch date to the user's desired date. Verse 2 tersely directs the user to perform this computation 'as stated in the handbook', i.e., the *Karaṇakutūhala*, and then to break down the result as follows:

> *kṛtvādau karaṇoktavāsaragaṇaṃ śiṣṭaiḥ suhṛṣṭātmabhir*
> *bhājyaṃ khāgni 30 mitair avāptakam idaṃ sūryair 12 vibhājyaṃ punaḥ* ||
> *labdhaṃ viṃśati 20 bhir bhajed atha catuḥśeṣāṅkasaṃjñā dhruvaṃ*
> *aṅkās te militāḥ svakoṣṭhakagatā laṅkānagaryāṃ khagāḥ* || 2 ||
>
> *śārdūlavikrīḍita*

Verse 2

> Firstly, computing the number of accumulated days as stated in the handbook, the learned who are cheerful in nature are to divide [it] by the amount 30; again this quotient should be divided by 12; one should divide the result by 20. Now, precisely these numbers called the four remainder-numbers, [entered into] their respective tables [with the corresponding entries] combined, are the [mean longitudes of the] planets at the city of Laṅkā [i.e., for zero degrees of terrestrial longitude].

As explained in Appendix, section 2.1, the algorithm stated in the *Karaṇakutūhala* for finding the accumulated civil days or *ahargaṇa*, here denoted c, uses the number y of integer calendar years since the epoch of Śaka 1105 (1183 CE), the number m_0 of integer calendar (synodic) months since the start of the current year, and the number t_0 of integer *tithis* (thirtieths of a mean synodic month) since the start of the current month. According to the *Brahmatulyasāraṇī* verse, this *ahargaṇa* c is to be divided by 30 to produce an integer number of completed (ideal) 'months' of 30 days each and a remainder D in days. That number of 'months' divided by 12 in turn yields an integer number of ideal 'years' of 360 days each and a remainder M in 'months'. The number of 'years' divided by 20 gives the number T of elapsed 20-'year' periods and a remainder Y in 'years'.[4]

4 For a fuller discussion of the *Brahmatulyasāraṇī*'s *ahargaṇa* procedure, *vide* (Montelle and

Presumably, most users of the *Brahmatulyasāraṇī* began their calculations by carrying out some form of this algorithm to determine the desired *c* and its partition into *T*, *Y*, *M* and *D*. However, one of our *Brahmatulyasāraṇī* manuscripts, MS Kh, provides an alternative method in the form of the tables described below.

Table XXV Accumulated civil days of the integer years since epoch in MS Kh.

This table lists the accumulated civil days corresponding to 30-year intervals and single luni-solar years.

ARGUMENTS:
(a) TABLE OF 30-YEAR INTERVALS: Year numbers from Śaka 1470 to 1740 (1548–1818 CE).
(b) TABLE OF SINGLE YEARS: 1 to 30 luni-solar years.

TABLE ENTRIES
(a) ROW 1: Civil days accumulated from epoch to the beginning of the numbered year. The first value represents the civil days accumulated from the epoch Śaka 1105 (1183 CE) to Śaka 1470 (1548 CE). The constant difference between entries is 10956 civil days, or 30 years assuming a year length of 365.2 days (365;12 sexagesimal).
(b) ROW 2: Civil days accumulated at the end of the numbered luni-solar year. The constant differences of 354 represent 29.5 civil days in a synodic month multiplied by 12. Occasionally the difference is 384 days which incorporates the periodic inclusion of a 30-day intercalary month. Only the entries for argument values 1 to 15 have been entered by the scribe.

Table XXVI Accumulated civil days of the solar months and solar days of the current year in MS Kh.

These tables are identified with the following paratext:

Plofker, 2015, vv. 2–3 on pp. 7–9). As noted therein, these mean motion tables with their idealized 'months' and 'years' in round numbers of civil days are in fact more reminiscent of some Islamic *zīj* calendar conversion tables than of the standard *ahargaṇa* algorithms in Sanskrit texts.

> *māsapraveśakṣepākapattraṃ* ‖ *dinapraveśakṣepakapattraṃ* ‖
>
> Page of offsets for the beginning of the [solar] month. Page of offsets for the beginning of the [solar] day.

They give the number of civil days to the beginning of the next solar month: that is, the number of days in each month it will take for the Sun to progress 30°, and the average time in days for the Sun to progress 1°, respectively.

ARGUMENTS:
(a) TABLE OF SOLAR MONTHS: 1 to 12 solar months.
(b) TABLE OF SOLAR DAYS: 1 to 12 solar months.

TABLE ENTRIES
(a) TABLE 1: Civil days required for the Sun to progress 30°. Function values are tabulated from an initial value of $30^d; 56, 55$ for argument 1 increasing to a maximum of $31^d; 37, 8$ at argument 3, decreasing to a minimum of $29^d; 19, 34$ at argument 9 and ending with a value of $30^d; 26, 27$ at argument 12. These values have been generated from the corresponding values in Table 2 multiplied by 30.
(b) TABLE 2: Average time in civil days it takes the Sun to progress 1° in the corresponding month. Function values are tabulated from an initial value of $1^d; 1, 43, 50$ for argument 1 increasing to a maximum of $1^d; 3, 14, 16$ at argument 3, decreasing to a minimum of $0^d; 58, 39, 8$ at argument 9 and ending with a value of $1^d; 0, 42, 0$ at argument 12.

These tables of *ahargaṇa* in MS Kh appear somewhat anomalous in the context of the *Brahmatulyasāraṇī*, as they employ a mean year length different from that of the Brāhmapakṣa, and solar days rather than *tithis* to measure the elapsed fraction of the current year. This is one of several indications that the *Brahmatulyasāraṇī*, at least in the form(s) available to us, was a somewhat eclectic set of procedures rather than a rigidly formalised canon.

3.2.2 *Planetary Epoch Mean Longitudes and Mean Longitude Increments since Epoch*

Once the *ahargaṇa* c from the epoch to the chosen date is obtained, by whatever method, the planetary mean longitudes for the chosen date at the end of c must be determined. As we discuss in Appendix, section 2.2, the term 'planet' in the context of mean longitudes in Indian astronomy embraces the Sun, the Moon, the lunar apogee, the lunar node, Mars, Mercury's *śīghra* or synodic apogee, Jupiter, Venus's *śīghra*, and Saturn. The mean longitudes of Mercury and Venus themselves are assumed equal to that of the Sun. (Since the lunar node's longitudinal motion is westward, we have denoted it with a negative sign in parentheses in Schemas 2–6, but have not tampered with the signs of nodal parameters as attested in manuscripts in Schemas 7–10.)

The *Karaṇakutūhala*'s approach to this computation requires adding to a planet's predetermined epoch mean longitude $\bar{\lambda}_E$ the amount of its mean longitudinal displacement corresponding to *c* (modulo 360°), including a secular correction that we here term the 'Brāhma-*bīja*'. While the *Karaṇakutūhala* employs the term *kṣepaka* 'additive increment' specifically to refer to $\bar{\lambda}_E$, paratext in manuscripts of the *Brahmatulyasāraṇī* may use this word instead for other amounts of mean longitudinal displacement, sometimes incorporating $\bar{\lambda}_E$.

Tables I–IX Mean longitudinal displacement of the planets

As opposed to the *Karaṇakutūhala*'s determination of mean longitudinal displacement via a linear function of a single integer *c* for each planet, the *Brahmatulyasāraṇī* uses the abovementioned partition of *c* into four components to serve as the arguments for its tables. Namely, each planet's table of mean longitudinal displacement contains four sub-tables giving the increments to its mean longitude for the desired numbers of days *D*, ideal 'months' *M*, ideal 'years' *Y*, and 20-'year' periods *T*, respectively:

- SUB-TABLE 1: Argument 0 to 30 days.
  [Planet]-*dinabhogāḥ*
  Advances in days of [Planet].
- SUB-TABLE 2: Argument 1 to 12 ideal months (of 30 days).
  [Planet]-*māsabhogāḥ*
  Advances in months of [Planet].
- SUB-TABLE 3: Argument 1 to 20 ideal years (of 360 days).
  [Planet]-*varṣāṇi*
  Years of [Planet].
- SUB-TABLE 4: Argument 1 to 30 periods of 20 ideal years.
  [Planet]-*kṣepakāḥ*
  Increments of [Planet].

The predetermined epoch mean longitude or 'offset' $\bar{\lambda}_E$ prescribed for each planet in the *Karaṇakutūhala* is incorporated into the first entry of the planet's 20-'year'-period table of mean longitudinal displacement (*vide* Schema 2).[5]

5 The *Karaṇakutūhala* epoch mean longitudes have also been stated separately in the paratext of this table in MS B.

SCHEMA 2 Epoch offsets from the *Karaṇakutūhala* included in the first entry in the 20-'year'-period table for each planet in the *Brahmatulyasāraṇī* tables of mean longitudinal displacement

Planet	$\bar{\lambda}_E$
Sun	10^s, 29°; 13, 0
Moon	10^s, 29°; 05, 50
Moon's apogee	4^s, 15°; 12, 59
Moon's node	(−)9^s, 17°; 25, 9
Mars	7^s, 21°; 24, 21
Mercury's *śīghra*	2^s, 21°; 14, 30
Jupiter	2^s, 4°; 0, 51
Venus' *śīghra*	8^s, 18°; 5, 55
Saturn	4^s, 3°; 43, 17

After entering into each of its tables with the appropriate value of D, M, Y or T, the user simply adds up the four corresponding table entries (modulo 360°) to get the total mean longitude for the body in question at the desired date.

The *Brahmatulyasāraṇī*'s tabulated values of mean longitudinal displacement $\Delta\bar{\lambda}$ corresponding to the number t of civil days in a single day, 'month', 'year', and 20-'year' period (i.e., 1, 30, 360, and 7200 civil days respectively) are listed for each planet in column 4 of Schemas 3–6. Column 2 in each Schema contains the corresponding value of $\Delta\bar{\lambda}$ reconstructed from the *kalpa*-parameters R and C (*vide* Appendix, section 2.2) by the relation

$$\Delta\bar{\lambda} = \frac{360\,R}{C} \cdot t \bmod 360$$

where t is the number of civil days in the time unit in question. Column 3 shows the slightly different value of $\Delta\bar{\lambda}$ reconstructed from Bhāskara's formulae in *Karaṇakutūhala* 1.7–12. The small discrepancies suggest that the *Brahmatulyasāraṇī* table entries were computed using the *Karaṇakutūhala* formulae instead of the simple *kalpa*-parameter ratios without Brāhma-*bījas*.

SCHEMA 3 Daily mean longitudinal displacements for the nine planets. Column 2: Reconstructed value (to arc-fifths) from *kalpa*-parameter ratio. Column 3: Reconstructed value (to arc-fifths) from *KKu* 1.7–12 formula. Column 4: Attested value (to arc-thirds) in the first entry in the daily table in Tables I–IX

Planet	$\Delta\bar{\lambda}^{(\circ)}$ **from *kalpa*-parameters**	$\Delta\bar{\lambda}^{(\circ)}$ **based on *KKu* 1.7–12**	$\Delta\bar{\lambda}^{(\circ)}$ **from Tables I–IX**
Sun	0; 59, 8, 10, 21, 33	0; 59, 8, 10, 12, 40	0; 59, 8, 10
Moon	13; 10, 34, 52, 46, 30	13; 10, 34, 52, 31, 50	13; 10, 34, 52
Moon's apogee	0; 6, 40, 53, 56, 32	0; 6, 40, 53, 50, 18	0; 6, 40, 53
Moon's node	(−)0; 3, 10, 48, 20, 6	(−)0; 3, 10, 48, 25, 15	(−)0; 3, 10, 48
Mars	0; 31, 26, 28, 6, 47	0; 31, 26, 28, 9, 45	0; 31, 26, 28
Mercury's *śīghra*	4; 5, 32, 18, 27, 45	4; 5, 32, 21, 1, 3	4; 5, 32, 21
Jupiter	0; 4, 59, 9, 8, 37	0; 4, 59, 8, 53, 59	0; 4, 59, 9
Venus' *śīghra*	1; 36, 7, 44, 35, 18	1; 36, 7, 43, 49, 49	1; 36, 7, 44
Saturn	0; 2, 0, 22, 51, 43	0; 2, 0, 23, 3, 34	0; 2, 0, 23

SCHEMA 4 Ideal 'monthly' mean longitudinal displacements for the nine planets. Column 2: Reconstructed value (to arc-fifths) from *kalpa*-parameter ratio. Column 3: Reconstructed value (to arc-fifths) from *KKu* 1.7–12 formula. Column 4: Attested value (to arc-thirds) in the first entry in the 'monthly' table in Tables I–IX

Planet	$\Delta\bar{\lambda}$ **from *kalpa*-parameters**	$\Delta\bar{\lambda}$ **based on *KKu* 1.7–12**	$\Delta\bar{\lambda}$ **from Tables I–IX**
Sun	0^s; 29°; 34, 5, 10, 46, 48	0^s, 29°; 34, 5, 6, 20, 24	0^s, 29°; 34, 5, 6
Moon	1^s; 5°; 17, 26, 23, 15, 6	1^s, 5°; 17, 26, 15, 55, 24	1^s, 5°; 17, 26, 16
Moon's apogee	0^s; 3°; 20, 26, 58, 16, 27	0^s, 3°; 20, 26, 55, 9, 16	0^s, 3°; 20, 26, 55
Moon's node	(−)0^s; 1°; 35, 24, 10, 3, 20	(−)0^s, 1°; 35, 24, 12, 37, 53	(−)0^s, 1°; 35, 24, 12
Mars	0^s; 15°; 43, 14, 3, 23, 52	0^s, 15°; 43, 14, 4, 52, 42	0^s, 15°; 43, 14, 4
Mercury's *śīghra*	4^s; 2°; 46, 9, 13, 52, 45	4^s, 2°; 46, 10, 30, 31, 43	4^s, 2°; 46, 10, 30
Jupiter	0^s; 2°; 29, 34, 34, 18, 41	0^s, 2°; 29, 34, 26, 59, 52	0^s, 2°; 29, 34, 27
Venus' *śīghra*	1^s; 48°; 3, 52, 17, 39, 13	1^s, 48°; 3, 51, 54, 54, 38	1^s, 48°; 3, 51, 55
Saturn	0^s; 1°; 0, 11, 25, 51, 58	0^s, 1°; 0, 11, 31, 47, 25	0^s, 1°; 0, 11, 33

3.3 Local and Secular Adjustments to Mean Longitudes

Preparing to correct the planetary mean longitudes to true longitudes corresponding to a particular location and time requires making geographical (*deśāntara*) and secular (*bīja*) adjustments to them. The secular correction terms in the *Karaṇakutūhala* that we call Brāhma-*bījas* have been discussed in the preceding section and more fully in Appendix, section 2.2. The next verse

SCHEMA 5 Ideal 'yearly' mean longitudinal displacements for the nine planets. Column 2: Reconstructed value (to arc-fifths) from *kalpa*-parameter ratio. Column 3: Reconstructed value (to arc-fifths) from *KKu* 1.7–12 formula. Column 4: Attested value (to arc-thirds) in the first entry in the 'yearly' table in Tables I–IX

Planet	$\Delta\bar{\lambda}$ **from *kalpa*-parameters**	$\Delta\bar{\lambda}$ **based on *KKu* 1.7–12**	$\Delta\bar{\lambda}$ **from Tables I–IX**
Sun	11^s, 24°; 49, 2, 9, 21, 4	11^s, 24°; 49, 1, 16, 4, 55	11^s, 24°; 49, 1, 16
Moon	2^s, 3°; 29, 16, 39, 1, 23	2^s, 3°; 29, 15, 11, 4, 50	2^s, 3°; 29, 15, 11
Moon's apogee	1^s, 10°; 5, 23, 39, 17, 27	1^s, 10°; 5, 23, 1, 51, 15	1^s, 10°; 5, 23, 1
Moon's node	(−)0^s, 19°; 4, 50, 0, 40, 8	(−)0^s, 19°; 4, 50, 31, 34, 44	(−)0^s, 19°; 4, 50, 33
Mars	6^s, 8°; 38, 48, 40, 46, 35	6^s, 8°; 38, 48, 58, 32, 24	6^s, 8°; 38, 48, 58
Mercury's *śīghra*	1^s, 3°; 13, 50, 46, 33, 8	1^s, 3°; 14, 6, 6, 20, 43	1^s, 3°; 14, 6, 6
Jupiter	0^s, 29°; 54, 54, 51, 44, 22	0^s, 29°; 54, 53, 23, 58, 28	0^s, 29°; 54, 53, 24
Venus' *śīghra*	7^s, 6°; 46, 27, 31, 50, 47	7^s, 6°; 46, 22, 58, 55, 44	7^s, 6°; 46, 22, 59
Saturn	0^s, 12°; 2, 17, 10, 23, 40	0^s, 12°; 2, 18, 21, 29, 2	0^s, 12°; 2, 18, 21

SCHEMA 6 Ideal 20-'year'-period mean longitudinal displacements for the nine planets. Column 2: Reconstructed value (to arc-fifths) from *kalpa*-parameter ratio. Column 3: Reconstructed value (to arc-fifths) from *KKu* 1.7–12 formula. Column 4: Attested value (to arc-thirds) in the first entry in the 20-'year'-period table in Tables I–IX

Planet	$\Delta\bar{\lambda}$ **from *kalpa*-parameters**	$\Delta\bar{\lambda}$ **based on *KKu* 1.7–12**	$\Delta\bar{\lambda}$ **from Tables I–IX**
Sun	8^s, 16°; 20, 43, 7, 1, 27	8^s, 16°; 20, 25, 21, 38, 24	8^s, 16°; 20, 25, 21
Moon	6^s, 9°; 45, 33, 0, 27, 44	6^s, 9°; 45, 3, 41, 36, 51	6^s, 9°; 45, 3, 41
Moon's apogee	2^s, 21°; 47, 53, 5, 49, 2	2^s, 21°; 47, 40, 37, 5, 19	2^s, 21°; 47, 40, 36
Moon's node	(−)0^s, 21°; 36, 40, 13, 22, 42	(−)0^s, 21°; 36, 50, 31, 34, 44	(−)0^s, 21°; 36, 50, 31
Mars	5^s, 22°; 56, 13, 35, 31, 57	5^s, 22°; 56, 19, 30, 48, 11	5^s, 22°; 56, 19, 30
Mercury's *śīghra*	10^s, 4°; 36, 55, 31, 2, 47	10^s, 4°; 42, 2, 6, 54, 25	10^s, 4°; 42, 2, 7
Jupiter	7^s, 28°; 18, 17, 14, 47, 28	7^s, 28°; 17, 47, 59, 29, 20	7^s, 28°; 17, 47, 59
Venus' *śīghra*	0^s, 15°; 29, 10, 36, 55, 55	0^s, 15°; 27, 39, 38, 34, 48	0^s, 15°; 27, 39, 38
Saturn	8^s, 0°; 45, 43, 27, 53, 24	8^s, 0°; 46, 7, 9, 40, 43	8^s, 0°; 46, 7, 9

of the *Brahmatulyasāraṇī* instructions briefly (and not entirely clearly) alludes to some other corrections:

madhyāḥ svadeśīyakhagā bhaveyur
deśāntareṇābdabhavātharāma- ‖
bījena yuktā gaṇakais tataś ca
spaṣṭāḥ kriyante phalayugmakena ‖ 3 ‖ *indravajrā*

VERSE 3

> The mean [longitudes] should become [longitudes of] the planets for one's own locality [when] adjusted by the *deśāntara*-correction [and] by the annual *rāmabīja* [correction (or, by the *rāmabīja* together with the annual correction?)]. And from [those], the true [longitudes] are made by the calculators by means of the two equations [i.e., *manda* and *śīghra* equations].

With the exception of one manuscript (see section 3.3.4), our *Brahmatulya-sāraṇī* sources do not include tables specifically devoted to these corrections. But as we describe in the following sections, two other manuscripts state a few values for them in the paratext of Tables I–IX.

3.3.1 *The Longitudinal-Difference Correction or* deśāntara

The so-called *deśāntara* correction serves to adjust the planetary mean longitudes for a locality with terrestrial longitude east or west of the earth's prime meridian, considered in Sanskrit astronomy to pass through Laṅkā on the earth's equator. *Karaṇakutūhala* 1.15 (*vide* Appendix, section 2.3) defines this correction in units of arcseconds by a formula equivalent to the following equation:

$$\textit{deśāntara} = \frac{\bar{v} \cdot d}{80},$$

where $\bar{v}$ is the planet's mean daily velocity or mean longitudinal increment $\Delta\bar{\lambda}$ for a single day, and d is the difference in terrestrial longitude between the given locality and the prime meridian, measured in units of *yojanas*.

It is presumably this formula that *Brahmatulyasāraṇī* verse 3 is prescribing for the *deśāntara*, but its computation is attested only by some *deśāntara* correction values rather haphazardly stated in table paratext in MSS SMB and S45 (reproduced in Schema 7). We have not been able to reconstruct from these apparently incoherent numbers any consistent value of the longitudinal difference d, although their positive sign indicates that the intended locality/ies should be west of the prime meridian.

SCHEMA 7 The *deśāntara* correction values stated in MSS SMB and S45 in paratext of Tables I–IX

Planet	Correction in MS SMB	Correction in MS S45
Sun		
Moon	$+ 296''$	
Moon's apogee	$+ 2''$	
Moon's node	$+ 1''$	
Mars	$+ 11''$	
Mercury's *śīghra*	$+ 92''$	
Jupiter	$+ 1'$	
Venus's *śīghra*		$+ 72''$
Saturn		0

3.3.2 *The Annual Correction or* abdabīja

As we interpret it, *Brahmatulyasāraṇī* verse 3 perhaps mentions in passing another correction to planetary mean longitudes also discussed by Bhāskara: namely, the *abdabīja* (literally 'annual correction'; *vide Karaṇakutūhala* 1.16, Appendix, section 2.4). These adjustments compensate for inaccuracies in the *Karaṇakutūhala*'s approximate mean-motion parameters for all the planets except the Sun, Mars and Saturn. The *abdabīja* for each of the remaining planets is computed by dividing the number y of elapsed years since the Śaka 1105 epoch by a specified divisor, and applying the (positive or negative) quotient in arcseconds to the planet's mean longitude.

MS S45 specifies numerical values for some of these *abdabīja* corrections, taking the number y of elapsed years since the epoch to be (approximately) 442, which appears to place the time of computation around 1625 CE. Schema 8 reproduces MS S45's values of y and *abdabīja*, along with *abdabīja* values recomputed by us for the given y using the *Karaṇakutūhala* parameters.

3.3.3 *The* rāmabīja *Corrections*

The so-called *rāmabīja* corrections to be added to or subtracted from the planetary mean longitudes appear to be a post-Bhāskara innovation attributed to one Rāma and frequently employed by astronomers in and after the sixteenth century (Pingree, 1996, pp. 168–171). These canonical values, and the versions of some of them appearing in MSS SMB and S45, are reproduced in Schema 9.

SCHEMA 8 The *abdabīja* corrections, in arcseconds, to planetary mean longitudes for stated numbers of elapsed years, as listed in MS S45 in paratext of Tables I–IX (columns 3–4), and as recomputed from the *Karaṇakutūhala abdabīja* factors (column 5)

Planet	*abdabīja* divisor (in *KKu* 1.16 and MS S45)	Elapsed years y (in MS S45)	*abdabīja* ('') (in MS S45)	*abdabīja* ('') (computed)
Moon	−78	442	−5; 46, 30, 0	5; 40
Moon's apogee	+30	442	−14; 45, 15	14; 44
Moon's node	+13	(omitted)		
Mercury	+22	442	+19	20; 5, 27...
		444	+20	20; 10, 55...
		446	+21	20; 16, 22...
Jupiter	−63	442	−6	7; 0, 57...
	(194 (?) in MS S45)	44[4]	−7	7; 2, 51...
Venus	−9	432	+0; 46	48
	(omitted in MS S45)	441	+49; 45, 0, 50	49

SCHEMA 9 The *rāmabīja* corrections, in arcminutes, for the planetary mean longitudes. Column 2: As attributed to Rāma in Caṇḍīdāsa's commentary on the *Karaṇakutūhala* (Pingree, 1996, Table 9 on p. 169). Columns 3 and 4: As stated in MSS SMB and/or S45 respectively, in paratext of Tables I–IX

Planet	*rāmabīja* (')	In MS SMB (')	In MS S45 (')
Sun	+2	+2	
Moon	−15	−15	
Moon's apogee	−30	+30	
Moon's node	−30	−30	
Mars	+80	+80	
Mercury's *śīghra*	+700	+700	
Jupiter	−190	−190	
Venus' *śīghra*	−270	−270	−270
Saturn	+90		+30

SCHEMA 10 The *deśāntara* and *rāmabīja* corrections stated in MS Kh. Cf. Table XXVII

Planet	*deśāntara*-correction			*rāmabīja*-correction			
Sun	0°	0′	44″	+	0°	2′	0″
Moon	0°	5′	24″	−	0°	15′	0″
Moon's apogee	0°	0′	19″	+	0°	0′	30″
Moon's node	0°	0′	0″	−	0°	30′	0″
Mars	0°	0′	2″	+	1°	20′	0″
Mercury's *śīghra*	0°	1′	32″	+	11°	40′	0″
Jupiter	0°	1′	13″	−	3°	10′	10″
Venus's *śīghra*	0°	1′	41″	−	4°	30′	0″
Saturn	0°	0′	2″	+	1°	30′	0″

3.3.4 *The* deśāntara *and* rāmabīja *Corrections and Apogee Longitudes in MS Kh*

Table XXVII

As noted above, only one of the *Brahmatulyasāraṇī* manuscripts, MS Kh, includes values for these corrections in tabular form. We have reproduced its tabulated *deśāntara* values in Schema 10, but as with the Schema 7 values included in paratext in MSS SMB and S45, we have not been able to extract from them any consistent value for the longitudinal difference.

The same table in MS Kh also contains versions of the *rāmabīja* corrections (see Schema 10) that likewise differ slightly from those in Schema 9.

Finally, the same table also lists the longitudes of the planets' orbital apogees, which are required for the next computational step of correcting their mean longitudes, as discussed below in section 3.4; these values are reproduced there in Schema 11.

3.4 Computation and Application of the *manda*-Equation to Mean Longitude and Velocity for the Seven Planets

The end of *Brahmatulyasāraṇī* verse 3 above introduced the subject of correcting mean planetary longitudes to true ones by means of the 'two equations', *manda* 'slow' and *śīghra* 'fast'. The first part of this correction procedure

SCHEMA 11 The *manda*-apogee longitudes as listed in *KKu* 2.1 (column 2), tabulated in MS Kh (column 3), and stated in MS B in the paratext of the planets' *manda*-equation tables (column 4). Cf. Tables XXVII, X, and XII–XVI

Planet	*manda*-apogee longitude (*KKu* 2.1)	*manda*-apogee longitude (MS Kh)	*manda*-apogee longitude (MS B)
Sun	2^s 18°	2^s 18° $0'$ $0''$	2^s 18° $0'$ $0''$
Mars	4^s 8° $30'$	2^s 8° $30'$ $0''$	4^s 8° $30'$ $0''$
Mercury	7^s 15°	7^s 15° $0'$ $0''$	7^s 15° $0'$ $0''$
Jupiter	5^s 22° $30'$	5^s 12° $30'$ $0''$	5^s 22° $30'$ $0''$
Venus	2^s 21°	2^s 28° $0'$ $0''$	2^s 21° $0'$ $0''$
Saturn	8^s 21°	7^s 28° $0'$ $0''$	7^s 28° $0'$ $0''$

involves the *manda*-correction: *vide* Appendix, section 2.5 and (Montelle and Plofker, 2015, v. 5 on pp. 12–16). It applies to the so-called 'seven planets', i.e., the five star-planets along with the Sun and the Moon.

The fundamental information required to determine this correction using the *Brahmatulyasāraṇī* tables is the so-called *manda*-anomaly κ_M, which the *Brahmatulyasāraṇī* nowhere explains how to compute. The *manda*-anomaly is the difference in longitude between the mean planet $\bar{\lambda}$ (which in the case of Mercury and Venus occupies the same position as the mean Sun) and the apogee of its own *manda*-circle λ_{A_M}. In the case of the Moon, as already noted, the mean longitude of its quickly revolving *manda*-apogee is separately tabulated and computed as for a planet. The other *manda*-apogees are considered to be effectively motionless, with canonical mean longitudes stated in *Karaṇakutūhala* 2.1 (*vide* Schema 17 in Appendix, section 2.5). Most of our *Brahmatulyasāraṇī* manuscripts omit this information but MSS Kh and B include versions of it, reproduced in Schema 11.

Verses 4 and 5 of the *Brahmatulyasāraṇī* instructions explain how to enter with the *manda*-anomaly into the tables of the *manda*-equation, and how to apply the corresponding corrections to the mean planet's position and velocity (*vide* Appendix, section 2.7):

kendrasya doraṃśamitiś ca koṣṭhe
bhuktaṃ tadagraṃ parabhogyakaṃ ca ||
kalādikaṃ tadvivarāhataṃ tu
ṣaṣṭyuddhṛtaṃ bhuktakamānakena || 4 || *upajāti*

Verse 4

The amount in degrees of the arc of the [desired] anomaly (*kendra*) is [entered] in the table. [The table entry for the degree] before that is the 'elapsed' (*bhukta*) and [then] the following 'future' (*bhogya*). The arcminutes etc. [of the argument] are multiplied by the difference of those [i.e., the two table entries] and divided by sixty, [and the result increased] by the amount of the 'elapsed'.

yuktaṃ bhaven mandaphalaṃ grahāṇāṃ
svarṇaṃ kramān meṣatulādikendre ||
grahasya bhuktir vivarāhataṃ ca
ṣaṣṭyuddhṛtaṃ kendravaśād dhanarṇam || 5 || *upajāti*

Verse 5

The *manda*-equation (*mandaphala*) of the planets should be applied positively or negatively [to the mean longitude of the planet] when the anomaly (*kendra*) is in [the semicircle] beginning with Aries or Libra respectively [i.e., when the anomaly is between 0 and 180 or between 180 and 360 degrees]. The velocity of a planet [is *manda*-corrected as follows: the fractional part of the desired value of anomaly], multiplied by the difference [between successive entries in the table row of velocity-correction (*gatiphala*)] and divided by sixty, [is the increment to the appropriate tabulated *gatiphala* entry. The resulting *gatiphala* is applied to the mean daily velocity] positively or negatively according to [whether] the anomaly [is in quadrants II and III or quadrants IV and I respectively].

The *bhukta* or 'elapsed' value refers to the table entry for the integer degree immediately preceding the desired argument, and the *bhogya* or 'future' value to the entry for the degree immediately after it. The concise instructions prescribe scaling the fractional difference between the desired argument value and the next lower integer degree by the difference between the two neighbouring table entries to give the required increment for the interpolated function value:

$$\text{desired value} = \textit{bhukta} + \frac{\textit{bhogya} - \textit{bhukta}}{60} \cdot \begin{pmatrix}\text{fractional part of}\\ \text{desired argument}\end{pmatrix}.$$

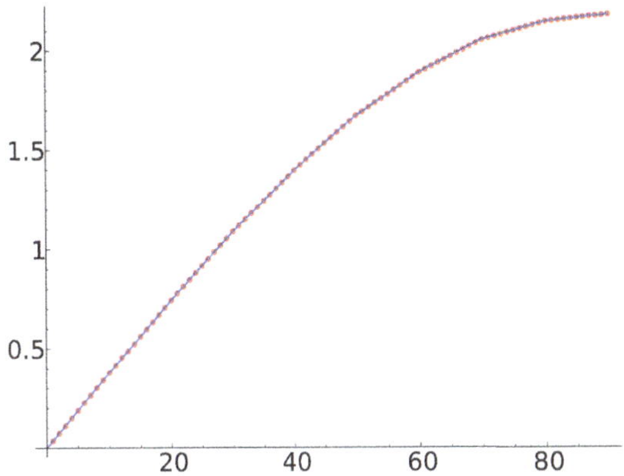

FIGURE 7 The Sun's *manda*-equation in degrees versus degrees of *manda*-anomaly, as reconstructed using its formula in *Karaṇakutūhala* 2.9–10 (line), and its tabulated values in *Brahmatulyasāraṇī* Table X (dots)

Tables X–XVI *Manda*-equation and velocity correction of the seven planets

The tables are labeled

> [Planet]-*mandaphalāni* ||

and their layout and content are summarized as follows:

ROW 1: Values of the *manda*-equation for argument 1 to 90 degrees of *manda*-anomaly. As the graph in figure 7 indicates for the case of the Sun, the table entries were computed using an algorithm equivalent to that given in *Karaṇakutūhala* 2.9–10. The function values were apparently calculated directly for every tenth entry, and the remaining entries derived by linear interpolation.

ROW 2: Successive differences of ROW 1.

ROW 3: Velocity correction (*gatiphala*) due to the *manda*-equation. These entries are derived from the rule given in *Karaṇakutūhala* 2.11cd–12, as the graph of a sample of the Sun's tabulated values in figure 8 shows. Like the *manda*-equation values, they have evidently been computed directly for every tenth entry.

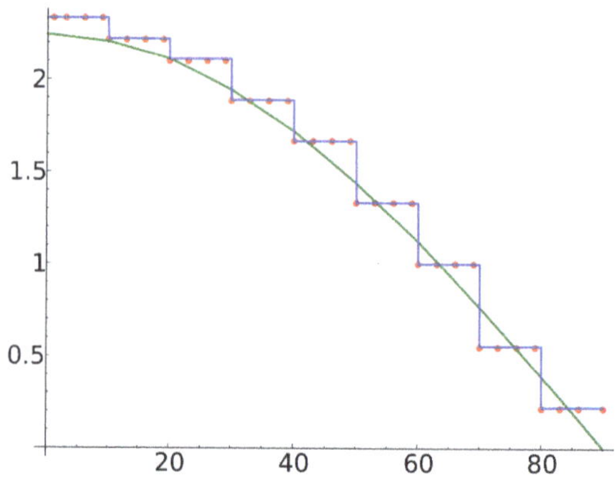

FIGURE 8 The Sun's *manda*-derived velocity correction in arcminutes versus degrees of *manda*-anomaly, as computed via *Karaṇakutūhala* 2.11–12 (stepwise line), and a sample of its tabulated values in *Brahmatulyasāraṇī* Table X (dots). The more accurate smooth version of the correction-term function from *Siddhāntaśiromaṇi* 2.36–38 (curved line) is shown for comparison; *vide* Appendix, section 2.7.

3.5 Computation and Application of the *śīghra*-Equation for the Five Planets; Completion of True Longitude and Velocity Corrections

3.5.1 *The* śīghra-*Correction to* manda-*Corrected Longitude; Iteration of Corrections*

While the Sun's and Moon's true longitudes are completely determined by the *manda*-correction, the five star-planets require the *śīghra*-correction as well; *vide* Appendix, section 2.8 and (Montelle and Plofker, 2015, vv. 6–7 on pp. 16–21). As explained therein, this correction requires knowledge of a planet's *śīghra*-anomaly, or the longitudinal difference κ_S between the planet's *manda*-corrected mean longitude λ_M and its *śīghra*-apogee longitude λ_{A_S}. The *śīghra*-apogee longitudes for Mercury and Venus are computed as discussed in section 3.2.2, while the mean Sun serves as the *śīghra*-apogee for the other three star-planets. The *śīghra*-correction ultimately depends also on the planet's own *para* or *śīghra*-circle radius (*vide* Schema 17 in Appendix, section 2.5), but these parameters are not explicitly stated in the *Brahmatulyasāraṇī* except in MS B, in the paratext of the *manda*-equation tables; cf. Tables XII–XVI.

Verses 6 and 7 of the *Brahmatulyasāraṇī* instructions explain how to determine the *śīghra*-anomaly, how to enter with it into the tables of the *śīghra*-equation, and how to apply the corresponding correction to the planet's *manda*-corrected longitude:

grahoṇam uccaṃ ca phalaṃ rasādhikaṃ
cet sūryataḥ śodhya lavādikaṃ kṛtam ||
bhāgāṅkasaṃkhyāgatakoṣṭhakaṃ tayoḥ
kalādikaṃ śeṣaṃ vivarāhataṃ tat[6] || 6 || *vaṃśamālā*

ṣaṣṭyā vibhaktaṃ svam ṛṇaṃ ca bhogyāt
kāryaṃ vihīnādhikatatkrameṇa ||
ādau hi mandārdhacalārdhakena
tasmāt samagranthapunaḥ punaś ca || 7 || *indravajrā*

Verses 6–7

[The longitude of] the apogee is diminished by [the longitude of] the planet. Having subtracted the result from 12 [signs] if it is greater than 6 [signs], it is made into degrees etc. [Subtract from this reduced *śīghra*-anomaly] the previous table entry for the number [equal to its] number of degrees; the remainder [from the subtraction] of those two is in arcminutes and so on. That is multiplied by the difference [between the previous and the next table entries, and] divided by sixty. [The result is] applied [to the previous entry] positively or negatively, according as that is respectively less or greater than the next entry. At first, [the mean longitude is corrected] with half the *manda*-equation and half the *śīghra*-equation, afterwards with the whole, repeatedly.

That is, because the absolute value of the *śīghra*-equation σ is symmetrical about 0° and 180° of anomaly κ_S, it is tabulated only for the first 180° of κ_S, so the user's given *śīghra*-anomaly must be reduced to the first two quadrants if it exceeds 180°. The appropriate increment to the tabulated value of σ for the previous integer degree of κ_S is found by linear interpolation. If the equation value is increasing from this table entry to the next, the increment must be added to the entry, or subtracted from it if σ is decreasing.

The verse does not explain how to determine the sign of the equation thus computed (in fact, the rule stated for the *manda*-equation in verse 5 that the equation should be positive when the anomaly is less than 180° and negative otherwise applies to the relation between the *śīghra*-equation and *śīghra*-

6 As pointed out in (Montelle and Plofker, 2015, p. 17), the last line of verse 6 is metrically incorrect.

anomaly as well). But it does prescribe an iteration of the *manda-* and *śīghra-*correction steps similar to that of *Karaṇakutūhala* 2.14 (*vide* Appendix, section 2.9), except that here the initial halving of the equations is apparently invoked for all the planets rather than for Mars alone.

Tables XVII-A–XXI-C *Śīghra*-equation of the five star-planets.

These tables are labeled

> **[Planet]-*śīghraphalāni*** ||

and contain the following data:

ROW 1: Values of the *śīghra*-equation σ for argument 1 to 180 degrees of *śīghra*-anomaly $\varkappa_S$, evidently computed using an algorithm equivalent to that given in *Karaṇakutūhala* 2.13 (*vide* Appendix, section 2.8 and figure 9). All tabulated values have been computed directly to degrees, arcminutes, and arcseconds, without interpolation.
ROW 2: Successive differences of ROW 1.
ROW 3: The *śīghra*-hypotenuse H_S or the distance from the earth to the true position of the planet, likewise computed by the procedure specified in *Karaṇakutūhala* 2.13. It is not quite clear why the compiler of the *Brahmatulyasāraṇī* tables bothered to tabulate the values of this quantity, as none of the rules specified in the text requires the user to employ it.

3.5.2 *Using the* śīghra*-Equation Tables to Correct Planetary Velocity; Iteration of Corrections*

Bhāskara's formulae in the *Siddhāntaśiromaṇi* and *Karaṇakutūhala* for computing the effect of the *śīghra*-equation σ on the planet's *manda*-corrected velocity v_M are discussed in Appendix, section 2.10. The *Brahmatulyasāraṇī* in verses 8–9 explains how to use the tables of σ to find this velocity correction:

> *drākkendrabhuktir vivareṇa nighnā*
> *ṣaṣṭyuddhṛtaṃ svaṃ ca phalasya vṛddhau* ||
> *hrāsa ṛṇaṃ mandagater grahāṇāṃ*
> *kṛtām iti syāt sphuṭakheṭabhuktiḥ* || 8 || *upajāti*
>
> *yadā na śuddhā tu vilomaśodhyā*
> *śeṣeṣu vakrā bhavatīha bhuktiḥ* ||

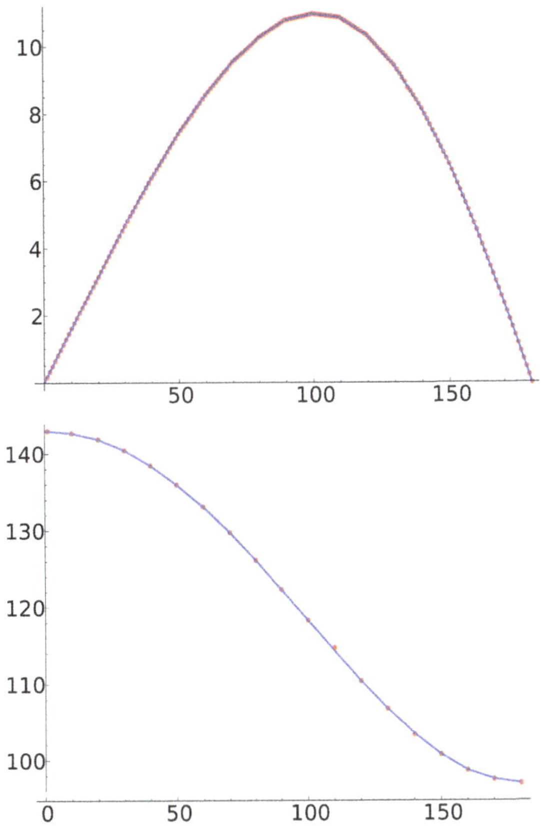

FIGURE 9 Functions for 0–180° of *śīghra*-anomaly for Jupiter, as represented by the *Karaṇakutūhala*'s algorithms (curve) and by sample values tabulated in *Brahmatulyasāraṇī* Table XIX-C (dots). **Above:** The *śīghra*-equation σ in degrees. **Below:** The *śīghra*-hypotenuse H_S in radial units ($R = 120$)

bhaumādikāḥ karmacatuṣṭayena
kujas tu yāvat sthiratām upeti ‖ 9 ‖ *upajāti*

VERSES 8–9

The velocity of the *śīghra*-anomaly is multiplied by the difference [between successive *śīghra*-equation values corresponding to that *śīghra*-anomaly] and the quotient with sixty [is applied] positively with respect to the *manda*[-corrected] velocity of the computed planets when there is increase of the equation [in successive tabulated values], negatively when there is decrease. Thus the velocity of the true planet should be [computed]. When [the modified *śīghra*-anomaly velocity] is not [capable of

> being] subtracted [from the *śīghra*-apogee velocity, it] is to be reverse-subtracted. The velocity here becomes retrograde in [the amount of] the remainders. The [star-planets] beginning with Mars [are corrected] by four procedures, but Mars [itself] until [it] attains fixedness.

That is, the planet's true velocity v in arcminutes per day is to be obtained from

$$v = v_M + v_{\varkappa_S} \cdot \frac{\Delta\,\sigma^{(\prime)}}{60},$$

where:
- v_M is the planet's mean velocity $\bar{v}$ corrected by the *manda*-velocity correction as per the rule in section 3.4;
- $v_{\varkappa_S}$ is the velocity of the *śīghra*-anomaly, defined in Appendix, section 2.10 (but nowhere in the *Brahmatulyasāraṇī* instructions) as $v_{\varkappa_S} = v_{A_S} - v_M$ where v_{A_S} is the velocity of the *śīghra*-apogee; and
- $\Delta\sigma$ is the difference (in arcminutes) between two successive tabulated values of the *śīghra*-equation σ.

(We take $\Delta\sigma$ to be positive when σ is increasing and negative when σ is decreasing, so we write '+' instead of '±' in the velocity formula.)

In Appendix, section 2.10 we explain the definition of v in modern mathematical terms as

$$v = v_M + v_{\varkappa_S} \cdot \frac{d}{d\varkappa_S}(\sigma) \quad = v_{A_S} - v_{\varkappa_S} \cdot \left(1 - \frac{d}{d\varkappa_S}(\sigma)\right).$$

Recalling that each difference $\Delta\sigma$ between successive σ-entries corresponds to one degree or 60 arcminutes of anomaly $\varkappa_S$, we can see that the first form of this definition is identical to the *Brahmatulyasāraṇī*'s formula for v, up to the equivalence of the derivative $\frac{d}{d\varkappa_S}(\sigma)$ with its finite-difference approximation $\Delta\sigma$. We likewise show that

$$v_M + v_{\varkappa_S} \cdot \frac{d}{d\varkappa_S}(\sigma) = v_M + v_{\varkappa_S}\left(1 - \frac{R\cos\sigma}{H_S}\right) = v_{A_S} - \frac{v_{\varkappa_S} \cdot R\cos\sigma}{H_S}$$

(where H_S as before denotes the *śīghra*-hypotenuse), which is Bhāskara's rule for v in *Siddhāntaśiromaṇi* 2.39. Thus the *Brahmatulyasāraṇī*'s v-rule might be described as a discrete version of the continuous function stated in the *Sid-*

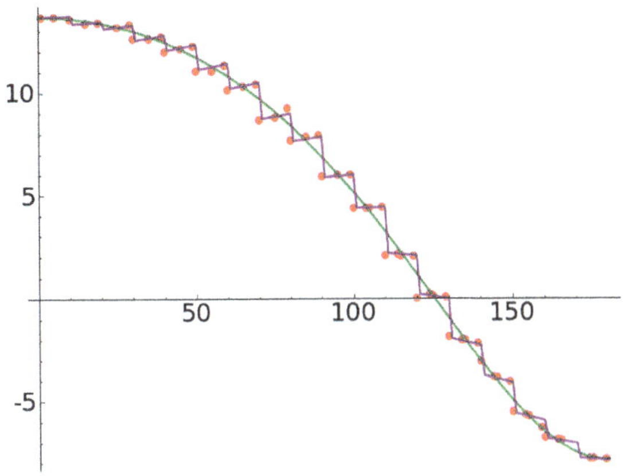

FIGURE 10 The *śīghra*-corrected true velocity in arcminutes per day, assuming $v_M = \bar{v}$, for 0–180° of *śīghra*-anomaly for Jupiter: The *Siddhāntaśiromaṇi* formula (smooth; using modern trig functions) compared to the *Brahmatulyasāraṇī* formula using σ-values computed in accordance with the *Brahmatulyasāraṇī*'s σ-rule (jagged curve), and the *Brahmatulyasāraṇī* formula using sample σ-values tabulated in Table XIX-C (dots)

dhāntaśiromaṇi; cf. the comparison of their results in the case of Jupiter in figure 10.[7]

Its instructions about correcting the velocity in the case of retrograde motion appear to refer to the subtractive rule(s) for true velocity v described in Appendix, section 2.10:

$$v = v_{A_s} - \frac{v_{\kappa_S} \cdot R\cos\sigma}{H_S} \quad (SiŚi\ 2.39), \text{ or}$$

$$v = v_{A_s} - \frac{v_{\kappa_S} \cdot R\,\text{sine-diff.}(\sigma)}{H_S} \cdot \frac{40}{7} \quad (KKu\ 2.17\text{ab}).$$

When the second term in either equation is larger than the first, the subtraction will produce a true velocity that is negative, i.e., retrograde. So the absolute value of this retrograde v will be found by 'reverse-subtracting' the first term from the second. Note that in the v equations, the factor by which v_{κ_S} is multiplied when subtracted from v_{A_S} in the retrograde-motion rule is not identical to—in fact, is the unit-complement of—the factor by which it is mul-

7 Note that a slightly modified version of this figure in (Montelle and Plofker, 2015, p. 25) misleadingly describes the pictured functions as '*śīghra* velocity correction' instead of '*śīghra*-corrected velocity'.

tiplied when added to v_M in the direct-motion rule. Thus the user needs to be careful about which form of the modified *śīghra*-anomaly velocity is required in this calculation.

At the end of verse 9 the issue of iterating the *manda* and *śīghra* correction sequence is raised again. Mars is supposed to be repeatedly corrected until the results for its true longitude converge on a fixed value, while two rounds of *manda* and *śīghra* are considered sufficient for the other planets.

3.5.3 *Using the* śīghra-*Anomaly of Mars to Correct Its* manda-*Apogee*

The *Karaṇakutūhala*'s application of a special procedure, characteristic of many Sanskrit treatises, for shifting the *manda*-apogee of Mars by an amount dependent on the planet's *śīghra*-anomaly is described in Appendix, section 2.6. The last of the *Brahmatulyasāraṇī* verse instructions explains the use of its table for carrying out this procedure; *vide* (Montelle and Plofker, 2015, v. 10 on pp. 26–28).

bhaumāśukendrasya padasya jāta-
gamyasya bhāgāḥ phalavat phalaṃ ca ||
kulīranakrādigate svakendre
hīnādhikaṃ spaṣṭam asṛṅmṛdūccam || 10 || *upajāti*

Verse 10

The degrees of the past [or] future [part, whichever is smaller,] of the quadrant of the *śīghra*-anomaly of Mars are like [the argument of an] equation [in the table of *manda*-apogee correction for Mars]. And the [corresponding] equation, when its own [*śīghra*-] anomaly is in Cancer or Capricorn, is [respectively] subtracted or added [to make] the *manda*-apogee of Mars accurate.

That is, the user enters into the table with the degrees of the smaller arc of the quadrant in which Mars' *śīghra*-anomaly falls, and finds in it the amount of the increment to be applied positively or negatively to Mars' *manda*-apogee.

Table XXII Correction for Mars' *manda*-apogee

bhaumamandoccasya spaṣṭīkaraṇārthaṃ koṣṭhakāḥ ||

The tabular entries for the sake of correcting the *manda*-apogee of Mars.

ROW 1: The numerical entries are generated by multiplying the table argument, which is the difference between the arc of the *śīghra*-anomaly of Mars and the closest integer multiple of 90°, by 3/20° or 9′. The function values are tabulated from a minimum value of 9′ for argument 1 to a maximum value of 6°40′ at argument 45.

ROW 2: Successive differences of Row 1. The entries are all 9′.

3.6 Corrections due to the Sun's Position

3.6.1 *Rising-Difference or* udayāntara *Corrections for the Sun and Moon*

This 'rising-difference' or *udayāntara* correction to solar and lunar longitudes as treated in *Karaṇakutūhala* 2.17cd–18 is explained in Appendix, section 2.11. The table entries in the *Brahmatulyasāraṇī* are based on the simple approximation algorithms described therein.

Tables XXIII-A–XXIV-C 'Rising-difference' correction for the Sun and the Moon.

atha ravicandrayor udayāntaraṃ koṣṭhakāḥ || *dvighnaṃ bhujāṃśopari* ||

Now, the tabular entries [providing] the 'rising-difference' (*udayāntara*) correction of the Sun and the Moon, [with its] above [argument] degrees of the arc [of longitude] multiplied by 2.

ROW 1: Values of *udayāntara*$_{\odot}$, the 'rising-difference' correction for the Sun, for argument 1 to 90 degrees (or in one case, every other degree from 2 to 90) of doubled solar tropical longitude $2\bar{\lambda}_{\odot}$. Function values are tabulated from a minimum value of 0 for argument 1 to a maximum value of 24″ for argument 90. The values have been computed directly for every tenth place and by linear interpolation for the remaining entries.

ROW 2: Values of *udayāntara*$_{☾}$, the 'rising-difference' correction for the Moon. Function values are tabulated from a minimum value of 0′ 6″

for argument 1 to a maximum value of 5′ 42″ for argument 90. The values have been computed directly for every tenth place and by linear interpolation for the remaining entries.

3.6.2 *Solar Declination*

For the trigonometric definition of this quantity, *vide* Appendix, section 2.12. One of our *Brahmatulyasāraṇī* manuscripts, MS Kh, includes the following table of its values.

Table XXVIII Solar declination

In this table, unlike in the *Karaṇakutūhala* rules mentioned in Appendix, section 2.12, the maximum declination value or ecliptic obliquity ϵ is taken as $1415' = 23^\circ\ 35'$. The table entries are labeled

krāntikalākoṣṭhakā ||

The tabular entries of the arcminutes of [solar] declination.

ROW 1: Values of the solar declination in arcminutes and arcseconds for argument 1 to 90 degrees of solar tropical longitude $\lambda_\odot$, based on the trigonometric definition of δ but apparently copied very carelessly. (*Vide* section 4.4 for a discussion of the scribe's practice of omitting significant digits.) Function values are tabulated from a minimum value of 24′ 0″ for argument 1 to a maximum value of 1415′ 0″ for argument 90.

ROW 2: Successive differences of Row 1. After argument 49 these are either zero or omitted.

CHAPTER 4

Variation in Manuscripts of *Brahmatulyasāraṇī* Tables

In light of the challenges described in section 1.3, we have had to articulate some criteria for significant as opposed to trivial distinctions between the manuscripts of the *Brahmatulyasāraṇī* used for this edition. The following (non-exhaustive) summary description attempts to organise, classify and justify these distinctions, using illustrative examples from the manuscripts. These examples are identified according to the roman numerals assigned to the *Brahmatulyasāraṇī* tables in Schema 1, and some of them are also accompanied by excerpts from images of manuscript pages.

It should be noted that manuscripts of other Sanskrit table texts also exhibit all the variations described here as well as additional ones (see, for instance, the many examples in (Montelle and Plofker, 2018)). So this chapter is by no means a comprehensive guide to the critical editing of texts in this genre. However, we hope that the following general observations will be useful in thinking about what such a guide might comprise.

Most fundamentally, we attempt to consistently distinguish a 'table' in the abstract, as a set of ordered n-tuples of function values corresponding to a sequence of argument values, from a row-and-column table grid as a concrete graphical object. This same sort of abstract/concrete distinction applies to individual symbols or numbers such as the zero, which in the abstract can represent a null value or empty place in a digit sequence, while its graphical form of a small open or filled circle can also signify an ellipsis in abbreviated Sanskrit words, a decorative element, etc.

4.1 Tables and Their Organisation

This section deals with manuscript variations in the individual table entities that are collected as a set in the *Brahmatulyasāraṇī*, the ways they are organised and ordered within it, and the numerical data they contain.

4.1.1 *Elements of the Table Set*

The selection of individual tables differs slightly in each of the manuscripts: *vide* Schema 1.

 DOI:10.1163/9789004432222_005

4.1.2 *Ordering of the Table Set*

Each of the manuscripts also orders some of its selected tables in a different way. For example:

- The tables of *udayāntara* values (Tables XXIII-A–XXIV-C) appear in the middle of MS S45 (f. 7[v]), at the beginning of MS S43 (f. 1[r]), and at the end of MS SMB (f. 5[v]).
- Some scribes (MSS B and S45) present each star-planet's table of *manda*-equation immediately followed by that for its *śīghra*-equation. Others (MSS Kh and S43) list first the *manda*-equation tables for all planets, and then all the corresponding *śīghra*-equation tables.
- The scribe of MS B reverses the usual ordering of the sub-tables of annual and vicennial increments (f. 4rv) in the table of longitudinal displacement of the lunar apogee (Table III), placing the latter before the former.

The planetary sequence for all tables pertaining to the five star-planets, however, is invariant. Namely, every scribe presents every type of planetary table in the so-called 'week-day' order of Sun, Moon, Mars, Mercury, Jupiter, Venus, Saturn.

4.1.3 *Inclusion in the Table Set of Tables from Other Works*

Some *Brahmatulyasāraṇī* manuscripts contain tables known or assumed to be borrowed from other table texts. The following examples are described in more detail in SUBS. 2:

- The numerical tables accompanying the *Brahmatulyasāraṇī* instructional verses in MS S29 are from the 16th-century *Candrārkī* of Dinakara rather than from the *Brahmatulyasāraṇī*.
- A note in MS Kh, following an incomplete table which does not appear to be part of the standard corpus of *Brahmatulyasāraṇī* tables, refers to the 1042 table text *Rājamṛgāṅka* of Bhojadeva.

4.1.4 *Combining Individual Tables*

Occasionally scribes will tabulate two separate functions together using the same argument sequence. For example:

- The table of the *udayāntara* correction for the Sun is combined with that for the Moon in MSS S45 and S43, *vide* figure 11a and figure 11b, using the same sequence of argument values. In contrast, the two functions are tabulated separately in MS SMB, *vide* figure 11c.

(a) Combined *udayāntara* tables for the Sun (रविः) and Moon (चंद्रः) with entries for all successive argument values (ंशा) 1–90 on MS S43 f. 1r. *Vide* Tables XXIII-A and XXIV-A

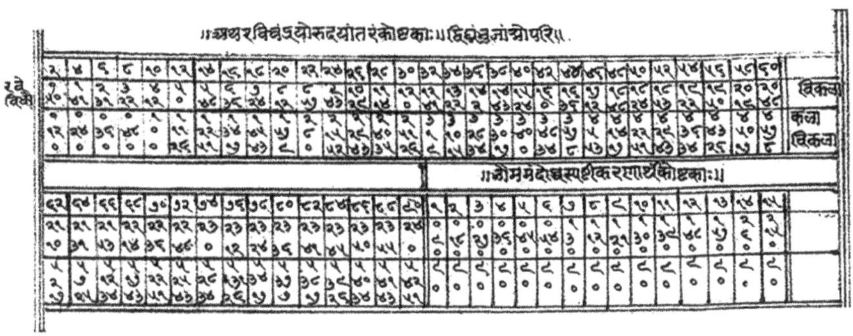

(b) Combined *udayāntara* tables for the Sun (रवे) and Moon (विधौ) with entries only for even values of the argument 2–90 on MS S45 f. 7v. *Vide* Tables XXIII-B and XXIV-B

(c) Separate *udayāntara* tables for the Sun (top) and Moon (bottom) with entries for all successive argument values 1–90 on MS SMB f. 5v. *Vide* Tables XXIII-C and XXIV-C

FIGURE 11 The *udayāntara* tables for the Sun and Moon from MSS S43 (f. 1r), S45 (f. 7v), and SMB (f. 5v)

4.1.5 *Incomplete Tables*

Tables may occasionally be left unfinished. For example:

- The scribe of MS Kh has failed to finish copying the row of successive differences in the solar declination table (Table XXVIII). An excerpt from the unfinished solar declination table on MS Kh f. 13[v] is shown below.

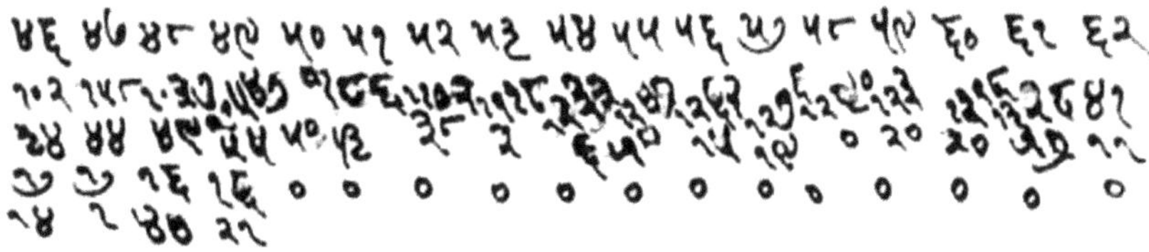

  After substituting zeroes for the actual difference values in several table cells, the scribe leaves the remaining cells blank.
- The scribe of MS Kh has not completed the table of accumulated civil days (*ahargaṇa*) for single years (Table XXV):

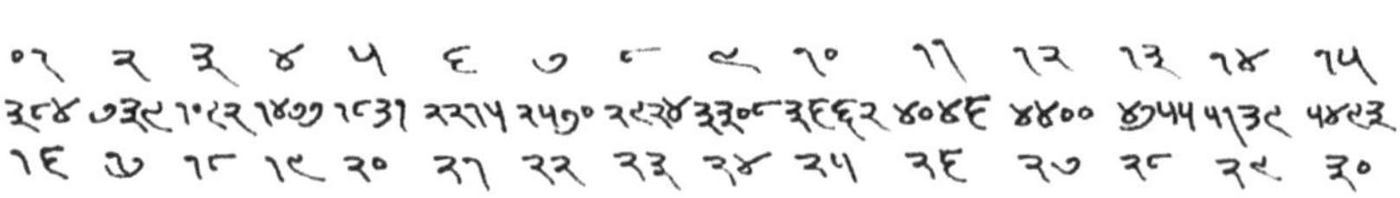

  The scribe has duly copied the table entries under the argument values 1–15 but omitted the ones for the argument values 16–30.

4.1.6 *Inclusion or Omission of Secondary Tabulated Functions*

Scribes may insert or omit table rows or columns containing specific kinds of numerical data, such as successive differences between table entry values. For example:

- The scribes of MSS B and Kh have left out the rows containing the differences between successive table entries in the tables of planetary *manda*-equations and *śīghra*-equations (Tables X–XXI-B).
- Some of the tables of mean longitudinal displacement in MS SMB include a column containing the constant difference between the entries in the table, as in the figure below:

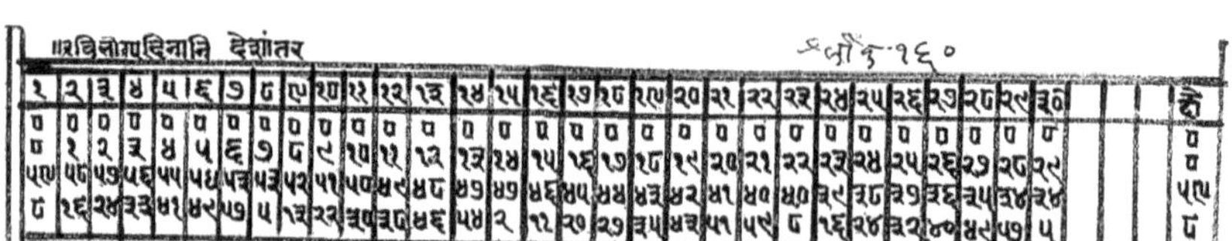

An extra column at the far right of this table of mean longitudinal displacement of the Sun (*vide* Table I) on f. 1ʳ of MS SMB contains the constant difference 0,0;59,8 between successive table entries.

4.1.7 *Additional or Omitted Table Entries*

The set of argument values for which entries are tabulated may change from one manuscript to another. For example:

- The tables of daily values of the mean longitudinal displacement of the planets (Tables I–IX) include entries for argument values 1–30 in MSS B, Kh, and SMB, and for argument values 0–29 (or in the case of the Moon, 1–29) in MS S45. The following excerpts from these tables for the Sun (left, f. 2ʳ) and the Moon (right, f. 2ᵛ) in MS S45 show their different initial argument values:

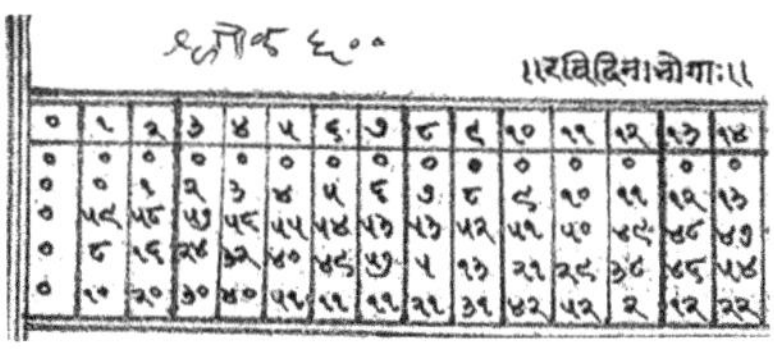

- The entries in the table of *udayāntara* for the Sun and Moon (Tables XXIII-B and XXIV-B) are listed only for even values of the argument in MS S45, *vide* figure 11b; whereas they are listed for both odd and even argument values in MSS S43 and SMB (Tables XXIII-A, XXIV-A, XXIII-C, and XXIV-C): *vide* figure 11a and figure 11c.

4.1.8 *Modified Epoch Offsets*

Scribes may change the values of additive offsets in units of time or longitudinal displacement to correspond to a chosen epoch date. For example:

- MS S45 (ff. 2ʳ–6ʳ) uses in its mean longitudinal displacement tables (Tables I–IX) a different set of vicennial argument values from those employed in MSS B (ff. 2ʳ–11ʳ) and SMB (ff. 1ʳ–5ʳ), implying an epoch date for MS S45 that is 600 years after the one used by MSS B and SMB.

4.1.9 *Numerical Values of Table Entries*

Different manuscript versions of what is nominally the same table may contain significantly different values of the table entries. For example:

- The *śīghra*-equation tables in MSS B (Tables XVII-A, XVIII-A, XIX-A, XX-A, and XXI-A) and Kh (Tables XVII-B, XVIII-B, XIX-B, XX-B, and XXI-B) differ substantially from their counterparts in MSS S43 and S45 (Tables XVII-C, XVIII-C, XIX-C, XX-C, XX-D, and XXI-C).
- The *śīghra*-equation table for Venus in MS S43 (Table XX-C) also differs from its counterpart in MS S45 (Table XX-D).

4.1.10 *Precision of Table Entries*

Different manuscript versions of the same table may differ in the precision or number of fractional sexagesimal places in their table entries. For example:

- The entries in the table of *udayāntara* values for the Sun and Moon (Tables XXIII-B and XXIV-B) in MS S45 (f. 7ᵛ) consist of integer and sexagesimal fractional arcseconds for the Sun and arcminutes with integer and sexagesimal fractional arcseconds for the Moon: *vide* figure 11b. The corresponding entries in MS S43 (f. 1ʳ), on the other hand, are precise only to integer arcseconds in both cases: *vide* figure 11a.
- The entries in mean longitudinal displacement tables (Tables I–IX) are given to the nearest arcsecond in MS SMB (ff. 1ʳ–5ʳ) but to the nearest arc-third (sixtieth of an arcsecond) in MS S45 (ff. 2ʳ–6ʳ) and MS Kh (ff. 1ʳ–5ʳ).
- The entries in the table row for the orbital hypotenuse (*karṇa*) in the *śīghra*-equation tables are precise only to the first sexagesimal fractional place in MS B: *vide* Tables XVII-A, XVIII-A, XIX-A, XX-A, and XXI-A, however, in MSS S43 and S45, they are precise to the second sexagesimal fractional place. *Vide* Tables XVII-C, XVIII-C, XIX-C, XX-C, XX-D, and XXI-C in our edition.

4.2 Paratext

Here we discuss different manuscript versions of the textual auxiliary material that appears within or around numerical tables.

4.2.1 *Table Headings/Titles*

Introductory text for a given table can vary between manuscripts from a one-word identification to an extended annotation. For example:

- The various versions of the heading for the table of *udayāntara* values (Tables XXIII-A–XXIV-C) include the following:
  - MS S43 f. 1ʳ reads (*vide* figure 11a):
    madhyamaṃ ravisāyanaṃ dvighnaṃ ca vidhāya bhujaḥ kāryas tadaṃśopari raver udayāntaraṃ vikalātmakaṃ tadadhaś candrasya kalādikaṃ sāyane madhyamaravau yugmapadasthe dhanaṃ ojapadasthe ṛṇaṃ | sāyanasūryasya dvighnasya dorjyā śarahṛd viliptā bhānoḥ vidhoḥ kvakṣi 21 hṛtā kalā sphuti udayāntaraṃ kāryam
    '[When one] has determined the precession-increased mean [longitude] of the Sun multiplied by two, a *bhuja* (i.e., an arc reduced to the first quadrant) is to be made; the rising-difference [correction] of the Sun [with its] superscribed [argument] in degrees of that [is given in] arcseconds (*vikalā*). Below that, the minutes and so on of the Moon. When the

precession-increased mean [longitude of the] Sun is situated in an even quadrant, it is positive; situated in an odd quadrant, it is negative. The sine of the precession-increased [mean longitude of the] Sun multiplied by 2, [in] arcseconds (*viliptā*) [when] divided by 5 for the Sun, [in] arcminutes (*kalā*) [when] divided by 21 for the Moon, is to be made [as] the accurate rising-difference [correction].'

- MS S45 f. 7ᵛ reads (*vide* figure 11b):
  atha ravicandrayor udayāntaraṃ koṣṭakāḥ dvighnaṃ bhujāṃśopari
  'Now, the tabular values [for] the rising-difference [correction] pertaining to the Sun and the Moon [with its] superscribed [argument] the degrees of the longitudinal arc multiplied by 2.'
- MS SMB f. 5ᵛ reads (*vide* figure 11c):
  atha dvighnasāyanaravibhujāśopari yātaphalaṃ kalādi | ojapade ṛṇaṃ | yugmapade dhanaṃ || atha candrasya yāntaphalaṃ vikalādiravivad dhanarṇaṃ
  'Now, the rising-difference [correction] (*yātaphala*) [with its] superscribed [argument], the arc (*bhuja*) of the precession-increased mean [longitude of the] Sun multiplied by two, is in minutes and so on. In an odd quadrant it is negative, in an even quadrant, positive. Now, the rising-difference of the Moon is positive or negative similarly to [that of] the Sun in arcseconds (*vikalā*) and so on.'

- The versions of the heading for the table of solar equation (Table X) include the following:
  - MS B f. 11ʳ reads:
    ravyādīnāmandaphalaṃ mandoccasūrya 2 | 18 | 0 | 0 |
    'The *manda*-equation of the Sun and so on. The solar apogee 2 | 18 | 0 | 0 |.'
  - MS S43 f. 2ʳ reads:
    raver mandakalāni || svarṇaṃ phalaṃ meṣatulādikendre ityanena sarvatra dharṇaṃ jñeyaṃ
    'The minutes of the *manda*-[equation] of the Sun. When the anomaly is in Aries and Libra the equation is positive or negative [respectively]. Thus everywhere it is to be understood as positive or negative [in this way].'
  - MS S45 f. 6ᵛ reads:
    mandaphalaṃ adho 'ntaraṃ tadadho gatiphalaṃ || ravimandaphalāni || adho gatiphalaṃ || ravimandoccaṃ 2 | 18 | 0 | 0 kendravaśād dhanarṇaṃ
    'The *manda*-equation. Below, the difference. Below that, the velocity correction. The *manda*-equations of the Sun. Below that, the velocity correction. The apogee of the Sun is 2 | 18 | 0 | 0. [The equation] is positive or negative depending on [the amount of] the anomaly.'

4.2.2 *Row Headers*

Headers identifying the content or units of the entries in a given row may appear, either on the left or right end of the row in question.

- The row headers for the tabulated *udayāntara* values for the Sun and Moon (Tables XXIII-B and XXIV-B) in MS S45 (*vide* figure 11b) contain the following information:
  - in the left margin, the identity of the planet to which the correction applies: top to bottom, *rave* 'to the Sun' and *vidhau* 'to the Moon'.
  - in the rightmost row cell, the units of the tabulated correction: (top to bottom) *vikalā* 'arcseconds' for the Sun, and *kalā* 'arcminutes' followed by *vikalā* 'arcseconds' for the Moon.
- The row headers in the rightmost row-cells of the table of the beginnings of solar days (Table XXVI) in MS Kh(a) f. 1v (excerpted image to the right) identify the following:
  - the argument: *dināntarāṇi* 'intervals in days'
  - the units of the tabulated entries: *vārāḥ* 'days' and successive sexagesimal fractions of days, *ghaṭī*, *palāni*, and *karāṇi*.

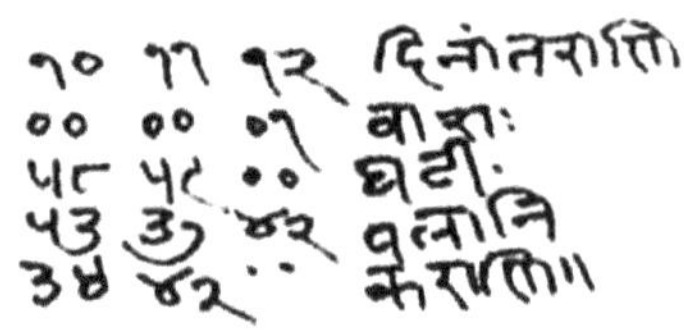

4.2.3 *Column Headers*

Individual columns occasionally appear with header text. For example:

- The constant-difference column entry shown in excerpted image to the right has the header *kṣe* for *kṣepaka* 'increment'.

4.2.4 *Table Colophons*

Sometimes scribes separate successive tables with colophon text. For example:

- MS Kh generally ends a table with the name of the tabulated function followed by *saṃpūrṇaṃ* 'complete', after which appears a brief heading for the next table consisting of the function identifier followed by *koṣṭakāḥ* ('table entries'), as in the excerpted image (MS Kh f. 13r) below:

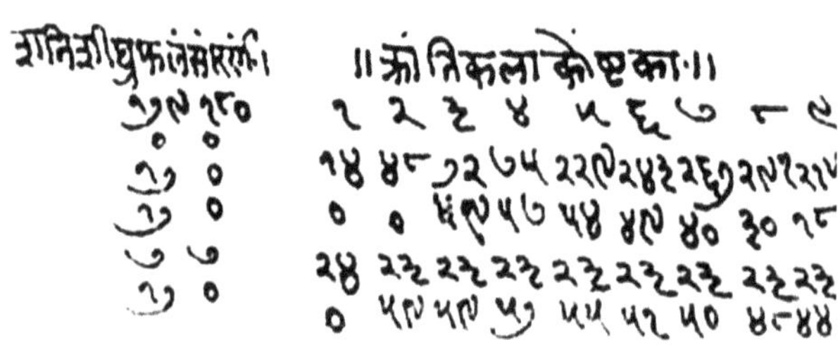

This image shows the end of the table of Saturn's *śīghra*-equation (Table XXI-B), followed by the table colophon *śaniśīghraphalaṃ saṃpūrṇaṃ* 'The *śīghra*-equation of Saturn is complete', and the start of its next table of solar declination values (Table XXVIII) with the header *krāntikalākoṣṭakāḥ* 'The table entries for the minutes of declination'.

– MS B uses a somewhat similar formula, with the table colophons ending in *samāptā* 'complete' and subsequent table headings beginning with *atha* 'now', as in the following excerpted image (MS B f. 20^{v}):

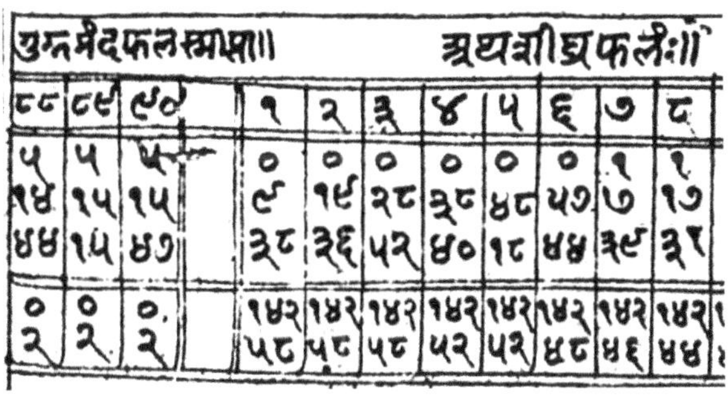

This excerpt shows the end of the table of Jupiter's *manda*-equation (Table XIV), followed by the table colophon *gurumandaphalas*[*a*]*māptā* 'The *manda*-equation [table] of Jupiter is completed', and the start of its next table for Jupiter's *śīghra*-equation (Table XIX-A), with the header *atha śīghraphalaṃ* 'Now, the *śīghra*-equation [table of Jupiter]'.

4.2.5 *Abbreviations/Morphology*

Sanskrit words or phrases in row/column headers and elsewhere in paratext may be written in abbreviated form, in stem form, or fully inflected. For example:

– The excerpted image (MS Kh f. 8^{v}) to the right shows abbreviations used in the table of epoch planetary apogee positions (Table XXVII) in MS Kh as row headers for the argument row (*gra*° for *graha* 'planet') and the units of the tabulated function (*rā* for *rāśī* 'zodiacal sign', *'ṃ* for *aṃśa* 'degree', *ka*° for *kalā* 'arcminute', and *vi* for *vikalā* 'arcsecond').

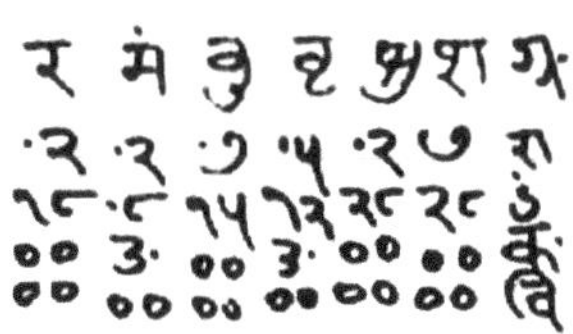

– Figure 11a shows the stem form *'ṃśā* 'degrees' in the row header identifying the units of the argument in the table of *udayāntara* values (Tables XXIII-A and XXIV-A) in MS S43.

4.2.6 *Notes and Annotations*

Marginal notes on any aspect of the topic may optionally be appended to a table, sometimes by the original scribe and sometimes apparently as annotations in a different hand. Such notes may be laid out in many different ways, as in the following examples:

- MS Kh f. 13^{v} shows marginal notes written at right angles to the table: *vide* figure 12a.
- MS S45 f. 6^{v} shows neatly written notes in the left margin: *vide* figure 12b.
- MS S45 f. 17^{v} shows table colophons in the top left margin and bottom right margins: *vide* figure 12c.

4.2.7 *Foliation and Running Titles*

Folio numbers are written on the folio verso, usually in the bottom right margin and sometimes also in the top left margin, where an abbreviation of the work's title may be written above them. They may also be accompanied by auspicious invocations. For example:

- At right, an excerpted image from MS Kh f. 1^{v} shows the folio number written in the bottom left corner along with an invocation of the name *śivaḥ* (which does not accompany the folio number on other pages).

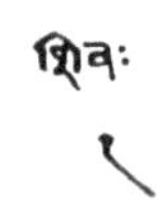

4.2.8 *Language*

Paratext, especially marginal notes, may be written in Sanskrit or in some (usually Sanskrit-related) vernacular. For example:

- MSS Kh f. 13^{v} and S45 f. 6^{v} show notes in some unidentified form of hybrid Sanskrit or Prakrit: *vide* figure 12a and figure 12b respectively.

4.3 Layout

In this section we treat manuscript characteristics relating to the spatial and visual features of tables.

4.3.1 *Page Orientation*

Manuscripts of numerical tables, like other Sanskrit manuscripts, are almost always written in landscape orientation, reflecting the original derivation of Indic manuscript formats from the ancient writing medium of longitudinally striated palm leaves. Consequently the tables' numerical data are tabulated horizontally, with the succession of argument values and their associated entries running from left to right across the page. Thus the 'row' element in San-

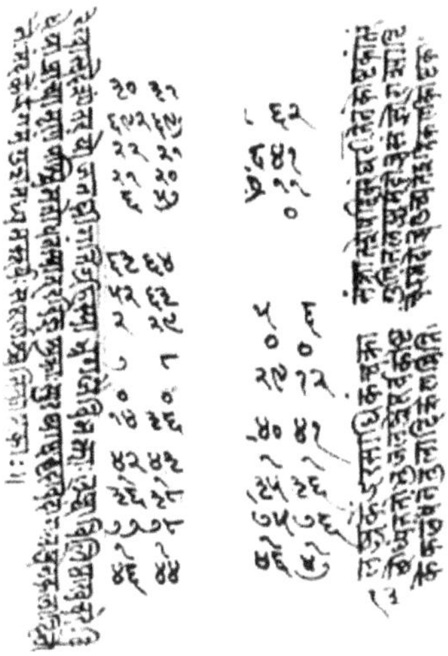

(left of f.13v) (right of f.13v)

(a) Excerpted image from f. 13[v] of MS Kh showing marginal text at a right angle to the tabulated values.

(b) Excerpted image from f. 6[v] of MS S45 showing written notes in the left margin.

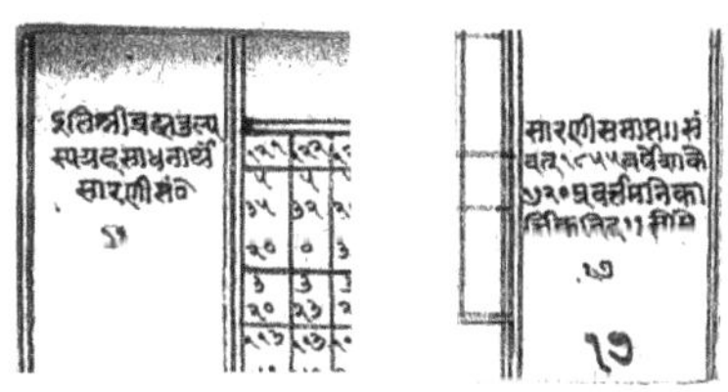

(top-left of f.17v) (bottom-right of f.17v)

(c) Excerpted image from f. 17[v] of MS S45 showing table colophons in the top left and bottom right margins.

FIGURE 12 Excerpts of folio images of the manuscripts of the *Brahmatulyasāraṇī* containing notes and annotations in various orientations

skrit tables typically corresponds to the 'column' element in their cuneiform, Greek, Arabic, Latin, etc., counterparts, and vice versa.

(a) Excerpt from the *śīghra*-equation table of Mars (Table XVII-C) with a single table wrap per page on MS S43 f. 6v

(b) Excerpt from the *śīghra*-equation table of Mars (Table XVII-C) with two table wraps per page on MS S43 f. 7r

FIGURE 13 Variable table wrapping per page in MS S43

4.3.2 *Breaking and Wrapping Long Tables*

A table containing too many successive entries to fit in a single row across a page is broken after the rightmost entry and wrapped underneath in a new table grid, likewise running from left to right. A long table may be written in several such table wraps stacked vertically, sometimes extending across multiple pages. The number of entries or table columns per table wrap may be uniform throughout the table, or may vary. For example:

- The table of the *śīghra*-equation of Mars (Table XVII-C) in MS S43 includes a single table wrap on f. 6v followed by two stacked table wraps on f. 7v, each containing exactly 30 columns: *vide* figure 13a and figure 13b.
- The entire table of the *manda*-equation of Mercury (Table XIII) in MS S45

(f. 10[r]) has been carefully fitted onto a single page in three stacked table wraps with 30 columns in each, as shown below.

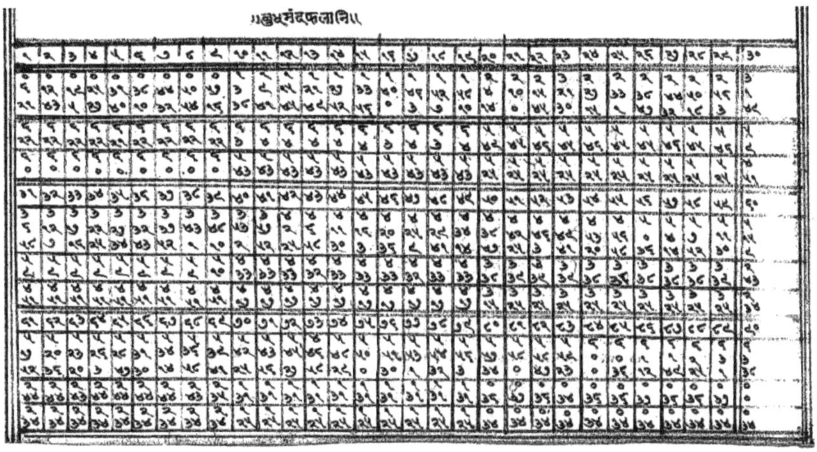

– The excerpted image below shows part of the table of the *śīghra*-equation of Mercury (Table XVIII-B) in MS Kh (f. 10[r]) containing three table wraps, each with a different number of columns.

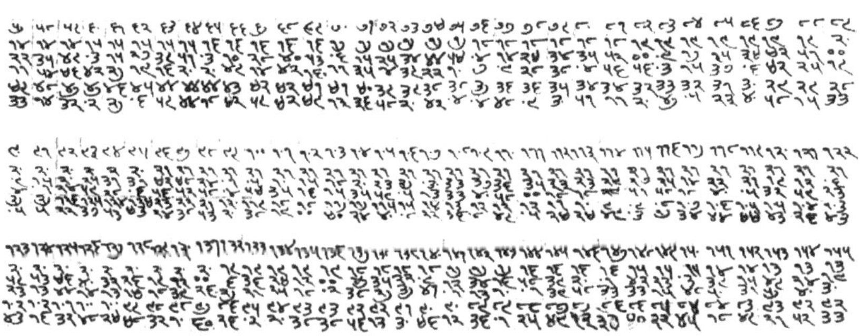

4.3.3 *Combining Tables in the Same Table Grid*

In general, scribes will lay out a separate table grid for each table, but exceptions are sometimes seen. For example:

– As the excerpted image below from MS B f. 16[v] shows, the transition from the *śīghra*-equation table of Mars (Table XVII-A) to the *manda*-equation table of Mercury (Table XIII) occurs within a single table grid. The scribe has ended the previous table with a double vertical rule and a single empty column before immediately beginning the next table with an initial double vertical rule in the same grid, which is then broken after argument value 14 and wrapped back to the left margin underneath.

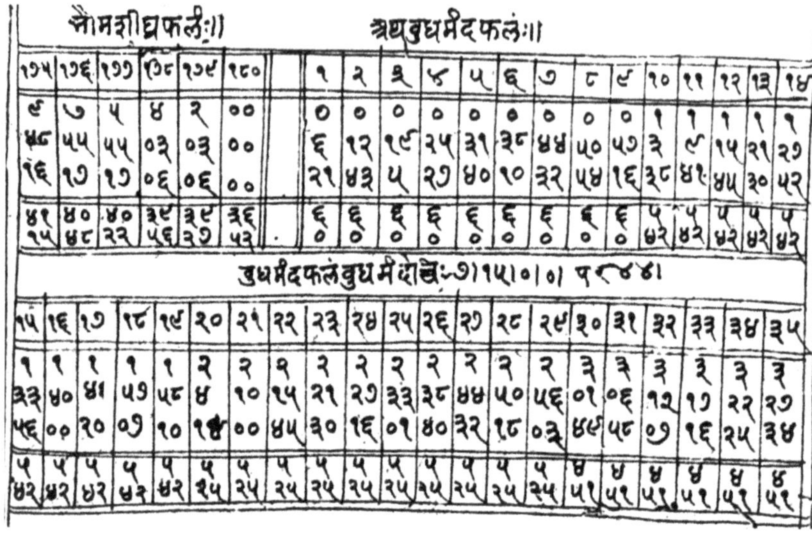

4.3.4 *Construction of Table Grids*

For the most part, table grids appear to have been pre-ruled by the scribe, and not always precisely tailored to the table dimensions, before filling in their numerical entries. The boundaries of table grids and/or the tables within them may be delineated by multiple vertical rules. For example:

- The tables of mean longitudinal displacement for Venus' *śīghra* (Table VIII) from MS SMB, an excerpt from which (f. 4^{v}) is shown at right, contain additional empty columns at the end, suggesting that the table grid was initially ruled without reference to the number of entries in the data.

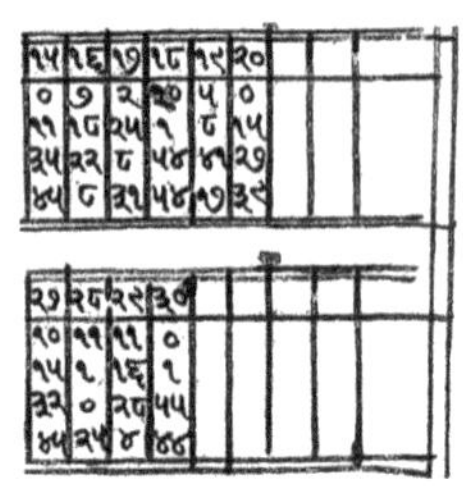

- The MS SMB tables of mean longitudinal displacement for Mars (Table V; see excerpt from f. 3^{r} at right) contain two final columns overhanging the right margin, evidently added by the scribe after it became clear that the originally ruled grid was inadequate to contain the table.

- MSS S43 and S45 generally use triple vertical rules to mark the left and right borders of a table grid. Excerpted images from f. 5^{v} of MS S43 (top) and f. 10^{v} of MS S45 (bottom) are shown below.

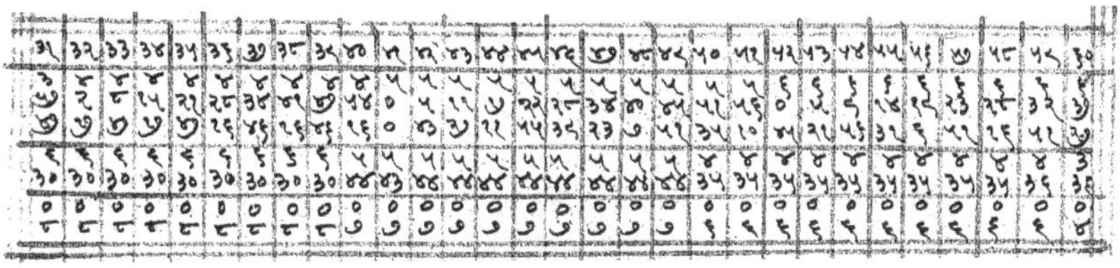

(f.5v of MS. S43)

(f.10v of MS. S45)

Double vertical rules (sometimes) serve to indicate the end or beginning of a table within a single table grid, e.g., as seen above at the end of the 30th argument on MS S45 f. 10^{v}.

– In MS B double horizontal rules are used to delimit the argument row from the entries beneath it: the image below shows an excerpt from f. 7^{v} of MS B.

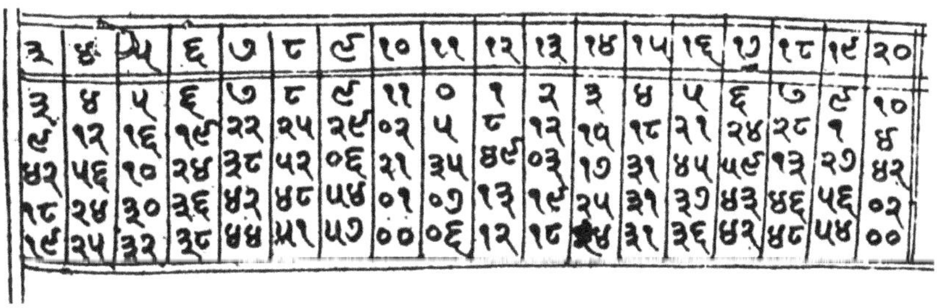

4.3.5 *Decorative Elements*

Some visual features of tables appear to be intended purely for decoration. For example:

– MS B on f. 25^{v} displays groups of zero-circles as a space filler in otherwise empty columns, as well as stylised florets. An excerpt from it is shown at right.

– MS Kh displays groups of zero-circles as a space filler in otherwise empty columns, presumably for aesthetic reasons. An excerpt from f. 2^{r} (bottom right corner) is shown at right.

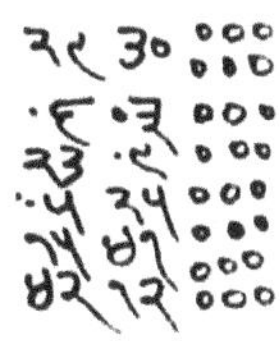

4.4 Representation of Numerical Data

The visual/symbolic representation of numerical content within tables likewise may vary from one manuscript to another. This section discusses several such variations.

4.4.1 *Null Values*

The presence of a null value in a table entry can be indicated by various uses of the zero symbol. For example:

- In MSS S43 and SMB a single zero is typically written in the appropriate cell: see figures 11a and 11c.
- In MS Kh a double zero is used for this purpose: see the excerpted image (from f. 2^{r}) at right. This practice, however, is not applied consistently throughout the manuscript.

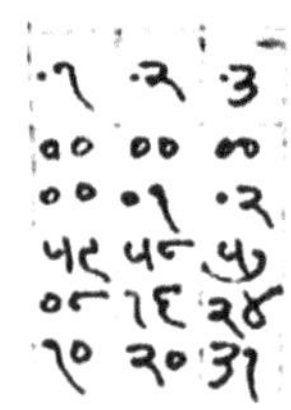

- In MS B both conventions are employed: see the excerpted image (from f. 16^{v}) at right.

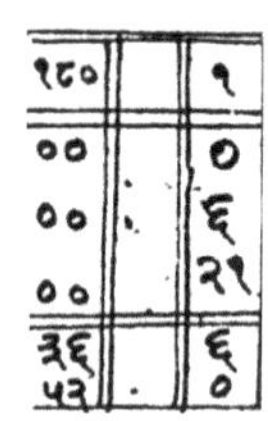

4.4.2 *Leading Zeros*

A zero symbol may be used to indicate the absence of more significant digits. For example:

- MS Kh frequently exhibits a zero-dot placed before a single digit: see the excerpted image (from f. 2^{r}) at right.

4.4.3 *Algebraic Sign Markers*

Some table cells include the symbols *dha* ('additive') and × ('subtractive') to indicate where values are subtractive/negative or additive/positive. The latter symbol derives from a short-hand for the *nāgarī* character ऋ or *ṛ*, an abbreviation of *ṛṇaṃ* meaning 'subtractive/negative' in Sanskrit.[1]

- MS Kh includes a row of sign markers in its table of *rāmabījas* (Table XXVII) on f. 8^{v}, shown at right.

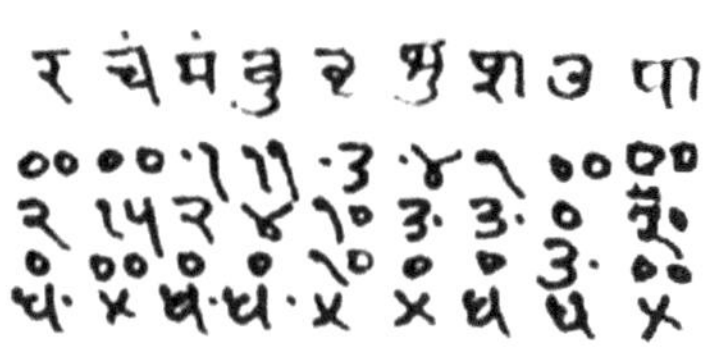

1 For further details and instances, see (Montelle and Plofker, 2018, 88 and *passim*).

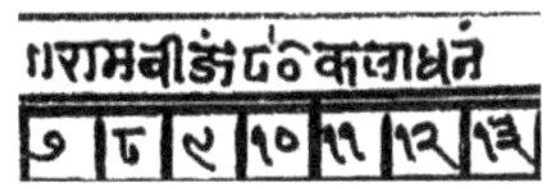

– MS SMB places the unabbreviated forms *dhanaṃ* and *ṛṇaṃ* in paratext after a number. The excerpted image at right (from f. 3r) shows the use of *dhanaṃ*.

4.4.4 *Omitting Repeated Digits*

When the most significant digits in a sequence of table entries remain unchanged throughout the sequence, scribes may opt to leave out the repeated digits, recording only the ones that change. For example:

– In the table of arcminutes of solar declination (Table XXVIII) in MS Kh, the first two (and sometimes even three) digits of the four-digit entries are frequently omitted unless they have changed from the previous entry. An excerpt from f. 13v is shown below, followed by a table listing its attested entries and their reconstructed full values.

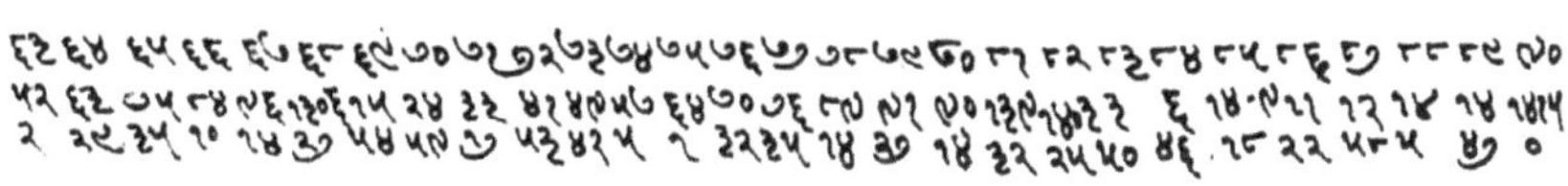

Argument	Presumed value	Attested reading	Argument	Presumed value	Attested reading
75	1364	64	82	1403	1403
76	1370	70	83	1403	3
77	1376	76	84	1406	6
78	1389	89	85	1409	1409
79	1391	91	86	1411	11
80	1390	90	87	1412	12
81	1391	1391			

The sequence of values in the arcminutes place attested in the manuscript for arguments 75–87 is shown in the third column; the corresponding numbers we presume the scribe intended to suggest are listed in the middle column.

4.4.5 *Numeral Forms*

Most forms of the *nāgarī*-script decimal place-value numerals used by scribes (*vide* the examples in Schema 12) are closely related and easily identifiable. We do not yet know enough about the slight differences in their palaeographic characteristics to draw definite conclusions about the chronological, regional or other information they may imply. *Vide*, e.g., (Sircar, 1965), (Salomon, 1998), and (Gokhale, 1967).

SCHEMA 12 Palæography of numeral forms seen in MSS B, Kh, S43, S45, and SMB

Numeral forms										
Modern Indo-Arabic	1	2	3	4	5	6	7	8	9	10
Modern *nāgarī*	१	२	३	४	५	६	७	८	९	१०
MS B (f. 5r)	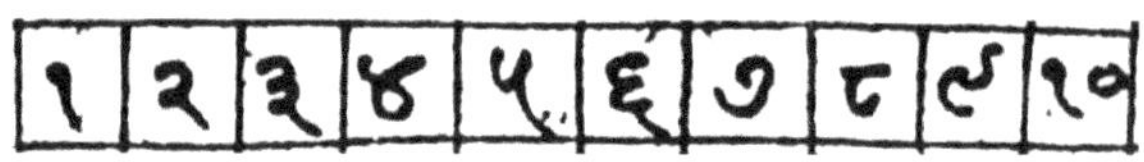									
MS Kh(b) (f. 1r)	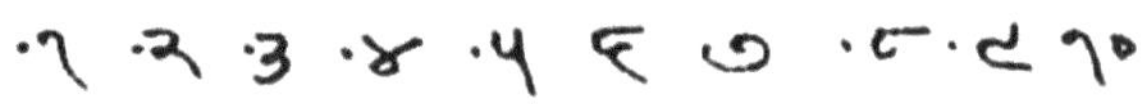									
MS S43 (f. 1r)										
MS S45 (f. 3v)										
MS SMB (f. 1r)	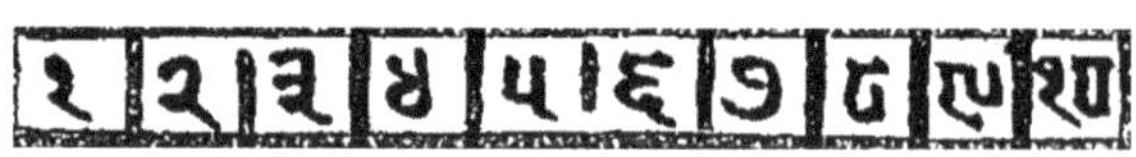									

CHAPTER 5

Framework and Features of the Critical Edition

Our critical edition of the instructional verses of the *Brahmatulyasāraṇī* is largely reproduced from (Montelle and Plofker, 2015). However, in the interim we have been able to consult an additional MS, Khasmohor 5424 (a,b), and have incorporated its readings into our apparatus as well as slightly modifying our translation.

We have constructed our critical edition of the *Brahmatulyasāraṇī*'s numerical tables to accommodate to the fullest possible extent the issues discussed in chapters 1 and 4. Adhering to the distinction between abstract and concrete interpretations of 'table' outlined at the start of chapter 4, we have chosen for the fundamental unit of the edited content the individual table rather than the manuscript folium. Each table is accompanied by its own critical apparatus describing its numerical and textual manuscript variants. At the same time, we have attempted to retain most of the original layout features of the tables as seen on the manuscripts, e.g., the vertical stacking of sexagesimal digits, dimensions of the table grids, placement of row headers and row breaks, and foliation breaks.

Section 5.1 lays out the basic typographic conventions employed in the edition of the *Brahmatulyasāraṇī*'s text and tables. In section 5.2, we outline our editorial choices specific to the tables edition in relation to more general principles of critical editing. Lastly, in section 5.3 we describe the visual structure we have assigned to the edited tables.

5.1 Typographic Conventions

The conventions described below are nearly identical to those we used in (Montelle and Plofker, 2015) and (Montelle and Plofker, 2013).

1. Scribal variants of *nāgarī* orthography which are emended silently and not noted in the critical apparatus (except where the meaning of the original reading may be ambiguous) include the following: a nasal consonant represented by *anusvāra* or a substituted incorrect nasal; omitted *visarga*, *virāma* or *avagraha*; misplaced *daṇḍas*; reversed conjunct consonants (e.g., *adba* for *abda*), or conjunct consonants that we cannot reproduce in our *nāgarī* typesetting; doubled consonants after *r* or across a *pāda* break; and routinely confused consonant pairs (e.g., *ba* for *va*, *ṣa* for *kha*) and all forms of *koṣṭa* for *koṣṭha* 'table entry'.

 DOI:10.1163/9789004432222_006

2. Fragments of Sanskrit words or compounds in *nāgarī* are indicated with a small circle ° at the breakpoint.
3. Folio breaks are indicated by '⌜' (beginning of the folio) and '⌟' (end of the folio). The end of a folio is only indicated for the edition of the tables.
4. In the critical apparatus, text followed by a single square close-bracket ']' indicates the edited version of the manuscript reading that follows it.
5. In the edited verse text, following the scribes of the two manuscripts, both *bhūtasaṃkhyā* numbers (word-numerals) and their corresponding digits are left in place.

5.2 Editing Problems and Editorial Choices for the Tables

5.2.1 *Location Identification within a Table*

Numerical tables have a built-in spatial structure for identifying the location of any given item within the content. This makes the usual editorial practice of specifying line numbers within the edited text somewhat superfluous. Consequently, we have instead identified individual table entries by their intrinsic spatial location. Rather than employing standard row-and-column coordinates for this purpose, we use an abbreviated key to designate the tabulated function allocated to a particular table row (including a sub-row affix to indicate the sexagesimal digit of the entry), together with the corresponding argument value of the function. This notational system is described in detail in section 5.3.

5.2.2 *Separate and Combined Versions of a Table*

Recensions or versions of a table that appear to have different computational methods are edited separately rather than being combined together into one. This is done to prevent the critical apparatus from being flooded with a vast number of variants for every entry, and also to acknowledge the computational differences that produce these different versions of the same table as separate entities. When the manuscripts use the same table grid and argument to combine multiple different functions, on the other hand, we assign each function to a separate table entity but preserve their combined spatial layout. See, for example, the combined tables of *udayāntara* values for the Sun and Moon in MS S43, *vide* Tables XXIII-A and Tables XXIV-A.

5.2.3 *Other Editorial Conventions*

5.2.3.1 Layout

The manuscripts' landscape format and overall layout of the tables are reproduced in the critical edition. The rulings of the original tables are also preserved as faithfully as possible to retain their visual style.

5.2.3.2 Script

Numerals in *nāgarī* have been transcribed to their modern Indo-Arabic equivalents. All paratext in *nāgarī* is transliterated to roman in the critical apparatus.

5.2.3.3 Morphology

Individual Sanskrit words or phrases that appear as row headers in the tables are given their inflected form in our edition, even though they vary inconsistently between inflected and stem forms in the different manuscripts.

5.2.3.4 Numerical Data Values

Our chief criterion for deciding what to present as the numerical values of the table entries themselves has been faithfulness to the manuscript witnesses. We have been very hesitant to discard an attested number in favour of a mathematically reconstructed one, unless it is obviously and unambiguously a mistake in a clear simple numerical sequence. And we have entirely refrained from 'correcting' mathematically unjustifiable readings at or beyond the third fractional place in a sexagesimal number. Authorial/scribal practices of rounding, truncating, interpolating, etc., are still so imperfectly understood that we balk at calling any anomalous value at that level of precision an 'error' in the historical text, however blatantly erroneous it may be in terms of modern mathematical recomputation.

5.3 Intrinsic Structure of the Edited Tables

In the absence of an unambiguous direction of text progression in numerical tables (since such tables can be read both vertically and horizontally), we have identified edited table entries in the *Brahmatulyasāraṇī* according to the natural ordering of the table argument values, along with the order of the associated sub-unit of the function value. For each of the tables, abbreviated keys are used to designate its various functions tabulated row-wise, and sub-row suffixes are used to indicate the successive place-value digits of the function values. The abbreviated keys and suffixes corresponding to the table functions and entry digits are listed in Schema 13.

andaphalāni ‖
dhanarṇaṃ ‖

22	23	24	25	26	27	28	29	30		
0	0	0	0	0	0	1	1	1		EC.d
48	50	53	55	57	59	1	3	5		EC.m
52	56	1	5	9	14	18	22	27		EC.s
2	2	2	2	2	2	2	2	1		DF.m
4	5	4	4	5	4	4	5	51		DF.s
2	2	2	2	2	2	2	2	1		ID.m
6	6	6	6	6	6	6	6	53⌟	⌟ f. 11ʳ B	ID.s

EC.m(16) 16 Kh; EC.s(16) 34 S45; DF.s(16) 10 S45
.s(20) 44 S45 DF.s(23) 4 S43 DF.s(24) 5 S43
; DF.s(30) 5 S43; ID.m(30) 2 B DF.s(36) 51 S43
7 B DF.s(51) 19 S43 EC.s(52) 57 B EC.s(53)
3 EC.s(58) 51 B, 49 Kh, 51 S43; DF.s(58) 19 S43
23 S43 EC.s(65) 22 S43 EC.s(66) 2 Kh, 51 S43
43, S45 EC.s(72) 34 Kh DF.s(78) 33 S43, S45

FIGURE 14 Excerpt of the edited table (top) and critical apparatus (bottom) of the *manda*-equation of the Sun, Table X on pp. 106–107

To better understand this function-and-argument spatial designation of the tabulated entries, consider the following example. Figure 14 shows an excerpt from the table of the *manda*-equation of the Sun, Table X on p. 106, followed by an excerpt from its critical apparatus on p. 107. The abbreviated function keys are seen outside the right margin of the table, viz. EC.d (degrees of the *manda*-equation) or ID.m (arcminutes of the *manda* velocity correction).

All textual variants within the critical apparatus are recorded using this notation system. E.g., the entry 'EC.s(52) 57 B' refers to the number in the arc**seconds** place of the table entry for the **solar *manda*-equation** function that corresponds to argument number **52.** (The remaining information in this description tells us that this number of arcseconds is given as 57 in MS B, rather than the value 58 appearing in the edited text.)

The other salient features of this notation system for the critical apparatus comprise the following:

- For a given argument value, all variants of different functions or sub-unit are separated by semicolons, whereas all variants of the same sub-unit of the same function are separated by commas. For instance, in the figure 14 example, we observe 'EC.m(16) 16 Kh; EC.s(16) 34 S45; …', but 'DF.s(78) 33 S43, S45…'.
- All variants corresponding to different argument values are separated by a double quadrat space. For instance, in the figure 14 example, we find 'DF.s(23) 4 S43 DF.s(24) 5 S43…'.

- In presenting the variants in the critical apparatus, the ordering of the functions takes precedence over the ordering of the manuscripts (based on their sigla). In the figure 14 example, we find 'DF.s(30) 5 S43; ID.m(30) 2 B'. Here, despite MS B preceding MS S43 alphabetically, we order the variants corresponding to the function 'DF' before those corresponding to the function 'ID'.
- Any metadata of the table structure that indicate any irregularities or idiosyncrasies in the argument sequence and/or the tabulated data format are put alongside the critical apparatus. This information is recorded to help recreate the original table (in any of the MSS) in its entirety from the edited version seen here. E.g., in the critical apparatus of the *manda*-equation of Mars, Table XII on p. 110, we note that MS Kh (f. 6[v]) has the *manda*-equation values corresponding to the arguments 83 to 90 left-shifted and tabulated under the arguments 82 to 89.

SCHEMA 13 The list of abbreviated keys used in preparing the critical edition of the tables of the *Brahmatulyasāraṇī*. Abbreviated keys for the different functions and subunits of the tables of the *Brahmatulyasāraṇī* corresponding to the argument '#' enclosed in parentheses

Tables I–IX of mean longitudinal displacement	pp. 72–102	DM.z(#)	Daily motion in zodiacal signs
		DM.d(#)	Daily motion in degrees
		DM.m(#)	Daily motion in arcminutes
		DM.s(#)	Daily motion in arcseconds
		DM.t(#)	Daily motion in arc-thirds
		MM.z(#)	Monthly motion in zodiacal signs
		MM.d(#)	Monthly motion in degrees
		MM.m(#)	Monthly motion in arcminutes
		MM.s(#)	Monthly motion in arcseconds
		MM.t(#)	Monthly motion in arc-thirds
		YM.z(#)	Yearly motion in zodiacal signs
		YM.d(#)	Yearly motion in degrees
		YM.m(#)	Yearly motion in arcminutes
		YM.s(#)	Yearly motion in arcseconds
		YM.t(#)	Yearly motion in arc-thirds
		IM.z(#)	Incremental motion (20 year periods) in zodiacal signs
		IM.d(#)	Incremental motion (20 year periods) in degrees
		IM.m(#)	Incremental motion (20 year periods) in arcminutes
		IM.s(#)	Incremental motion (20 year periods) in arcseconds
		IM.t(#)	Incremental motion (20 year periods) in arc-thirds
Tables X–XVI of *manda*-equation	pp. 106–118	EC.d	*manda*-equation in degrees
		EC.m(#)	*manda*-equation in arcminutes
		EC.s(#)	*manda*-equation in arcseconds

SCHEMA 13 The list of abbreviated keys (*cont.*)

		DF.m(#)	Successive differences in arcminutes
		DF.s(#)	Successive differences in arcseconds
		ID.m(#)	Velocity correction due to the *manda*-equation (*gatiphala*) in arcminutes
		ID.s(#)	Velocity correction due to the *manda*-equation (*gatiphala*) in arcseconds
Tables XVII-A–XXI-C of *śīghra*-equation	pp. 120–156	EJ.d(#)	*śīghra*-equation in degrees
		EJ.m(#)	*śīghra*-equation in arcminutes
		EJ.s(#)	*śīghra*-equation in arcseconds
		DF.m(#)	Successive difference in arcminutes
		DF.s(#)	Successive difference in arcseconds
		AH.a(#)	Hypotenuse (*śīghra-karṇa*) measure in sexagesimal units
		AH.b(#)	Hypotenuse (*śīghra-karṇa*) measure in sexagesimal sixtieth-parts
		AH.c(#)	Hypotenuse (*śīghra-karṇa*) measure in sexagesimal sixtieth-squared-parts
Table XXII of correction for Mars' *manda*-apogee	p. 159	MP.d(#)	Correction to Mars' *manda*-apogee in degrees
		MP.m(#)	Correction to Mars' apogee in arcminutes
		MP.s(#)	Correction to Mars' apogee in arcseconds
		DF.m(#)	Successive difference in arcminutes
		DF.s(#)	Successive difference in arcseconds
Table XXVIII of solar declination	p. 165	SL.m(#)	Solar declination in arcminutes
		SL.s(#)	Solar declination in arcseconds
		DF.m(#)	Successive difference in arcminutes
		DF.s(#)	Successive difference in arcseconds

CHAPTER 6

Critical Edition of Versified Text and Tables

6.1 Critical Edition of the Verses

⌜ओं श्रीगणेशाय नमः ॥ f. 6r S29
f. 1r Kh(a)

नत्वा वल्लभनन्दनं तदनुगोपालांह्रिपद्मद्वयं
ज्ञात्वा श्रीगुरुवाक्यतो ह्यहर्निशं सद्युक्तिमेवाधुना ।
सिद्धान्तेषु यथोक्तखेचरविधिः सुस्पष्टकोष्ठं मुहु-
र्मध्यस्पष्टविभागतो ग्रहगणात्कुर्वे दिनौघादहम् ॥ १ ॥

कृत्वादौ करणोक्तवासरगणं शिष्टैः सुहृष्टात्मभि-
र्भाज्यं खाग्नि ३० मितैरवाप्तकमिदं सुर्यै १२ र्विभाज्यं पुनः ।
लब्धं विंशति २० भिर्भजेदथ चतुःशेषाङ्कसंज्ञा ध्रुवं
अङ्कास्ते मिलिताः स्वकोष्ठकगता लङ्कानगर्यां खगाः ॥ २ ॥

मध्याः स्वदेशीयखगा भवेयु-
र्देशान्तरेणाब्दभवाथराम- ।
बीजेन युक्ता गणकैस्ततश्च
⌜स्पष्टाः क्रियन्ते फलयुग्मकेन ॥ ३ ॥ f. 6v S29

केन्द्रस्य दोरंशमितिश्च कोष्ठे
भुक्त तदग्र परभोग्यकं च ।
कलादिकं तद्विवराहतं तु
षष्ट्युद्धृतं भुक्तकमानकेन ॥ ४ ॥

युक्तं भवेन्मन्दफलं ग्रहाणां
स्वर्णं क्रमान्मेषतुलादि केन्द्रे ।
ग्रहस्य भुक्तिर्विवराहतं च
षष्ट्युद्धृतं केन्द्रवशाद्धनर्णम् ॥ ५ ॥

---

1 ओं] *om.* Kh   श्रीगणेशाय] श्रीसूर्याय Kh   3 ह्यहर्निशं] ह्यहनिशं Kh   सद्युक्तिमेवाधुना] द्युमेवाद्युना S29   4 सुस्पष्ट°] स्पष्ट° S29   कोष्ठं मु°] कोष्टोमु° S29;   कोष्टैर्मु° Kh   5 °गणात्कु°] °गणान् कु° Kh   ॥ १ ॥] *om.* Kh   7 °वाप्तक°] °वात्मक° S29   °कमिदं] °कमितं Kh   र्वि°] वि° S29   8 २०] *om.* Kh   °संज्ञा ध्रुवं] °संख्याः ध्रुवाः Kh   10–11 °युर्देशान्तरेणाब्दभवाथ°] °युदेशान्तरेणाप्रभथ° S29   °र्देशान्त°] °र्देषांत° Kh   14 °मितिश्च कोष्ठे] °मितेष्टकोष्टं Kh   कोष्ठे] कोष्टं S29   15 °ग्रं परभोग्यकं] °ग्रपरभोग्यंकं S29   तदग्रं] तदग्रे Kh   17 °मानकेन] °माप्तकेन Kh   20 °हतं]°हता Kh   21 °द्धृतं] °द्धृता Kh

 | DOI:10.1163/9789004432222_007

ग्रहोनमुद्यं च फलं रसाधिकं
चेत्सूर्यतः शोध्य लवादिकं कृतम् ।
भागाङ्कसंख्यागतकोष्ठकं तयोः
कलादिकं शेषं विवराहतं तत् ॥ ६ ॥

षष्ट्या विभक्तं स्वमृणं च भोग्या-
त्कार्यं विहीनाधिकतत्क्रमेण ।
आदौ हि मन्दार्धचलार्धकेन
तस्मात्समग्रन्थपुनः पुनश्च ॥ ७ ॥

द्राक्केन्द्रभुक्तिर्विवरेण निघ्ना
षष्ट्युद्धृतं स्वं च फलस्य वृद्धौ ।
ह्रास ऋणं मन्दगतेर्ग्रहाणां
कृतामिति स्यात् स्फुटखेटभुक्तिः ॥ ८ ॥

यदा न शुद्धा तु विलोमशोध्या
शेषेषु वक्राभवतीह भुक्तिः ।
भौमादिकाः कर्मचतुष्टयेन
कुजस्तु यावत्स्थिरतामुपेति ॥ ९ ॥

भौमाशुकेन्द्रस्य पदस्य जात-
गम्यस्य भागाः फलवत्फलं च ।
कुलीरनक्रादिगते स्वकेन्द्रे
हीनाधिकं स्पष्टमसृङ्मृदूद्यम् ॥ १० ॥

22 फलं] पलं Kh 25 °दिकं शेषं] °दि शेषं Kh 27 °धिकतत्क्र°] °धितत्क्र° S29 28–29 मन्दार्ध°…°पुनः] मन्दार्धकेन तस्मात्समग्रं पुनः S29 30 द्राक्केन्द्र°] द्रार्केन्द्र° Kh °भुक्तिर्वि°] °भुक्तिवि° S29 31 °द्धृतं] °द्धृता Kh फलस्य] पलस्य Kh 32 ह्रास ऋणं] ह्रासो ऋणं S29; ह्रासे ऋणं Kh मन्दगतेर्ग्र°] मन्दार्द्धगतेग्र° S29 °गतेर्ग्र°] °गतेग्र° Kh 33 °मिति] °सती Kh 35 वक्राभवतीह] भुक्तावहतीह Kh 36 भौमादिकाः] सौम्यादिका Kh 37 °मुपेति] °मुपैति S29, Kh 38 पदस्य] मंदस्य S29 38–39 जात°…°भागाः] यातगम्याल्पभागाः Kh 40 °रनक्रा°] °रक्रा° S29 41 °सृङ्मृदूद्यम्] °स्मृमृदुद्यं S29

6.2 Critical Edition of the Tables

Table I

⌜a f. 1^{r} Kh
f. 2^{r} S45
f. 1^{r} SMB
⌜b f. 2^{r} B

⌜a‖ ravidinabhogāḥ ‖

	0	1	2	3	4	5	6	7	8	9	10	11	12	13	14	15	16	17	18	19	20	21	22	⌜b23	24	25	26	27	28	29	30
DM.z	0	0	0	0	0	0	0	0	0	0	0	0	0	0	0	0	0	0	0	0	0	0	0	0	0	0	0	0	0	0	0
DM.d	0	0	1	2	3	4	5	6	7	8	9	10	11	12	13	14	15	16	17	18	19	20	21	22	23	24	25	26	27	28	29
DM.m	0	59	58	57	56	55	54	53	53	52	51	50	49	48	47	47	46	45	44	43	42	41	40	40	39	38	37	36	35	34	34
DM.s	0	8	16	24	32	40	49	57	5	13	21	29	38	46	54	2	10	18	27	35	43	51	59	7	16	24	32	40	48	56	5
DM.t	0	10	20	30	40	51	11	11	21	31	42	52	2	12	22	32	42	52	4	14	24	34	44	54	4	15	25	35	46	56	6

‖ ravimāsabhogāḥ ‖

	1	2	3	4	5	6	7	8	9	10	11	12
MM.z	0	1	2	3	4	5	6	7	8	9	10	11
MM.d	29	29	28	28	27	27	26	26	26	25	25	24
MM.m	34	8	42	16	50	24	58	32	6	40	14	49
MM.s	5	10	15	20	25	30	35	40	45	51	56	1
MM.t	6	13	19	25	32	38	44	51	57	3	10	16

‖ ravivarṣāṇi ‖

	1	2	3	4	5	6	7	8	9	10	11	12	13	14	15	16	17	18	19	20
YM.z	11	11	11	11	11	10	10	10	10	10	10	9	9	9	9	9	9	8	8	8
YM.d	24	19	14	9	4	28	23	18	13	8	2	27	22	17	12	7	1	26	21	16
YM.m	49	38	27	16	5	54	43	32	21	10	59	48	37	26	15	4	53	42	31	20
YM.s	1	2	3	5	6	7	8	10	11	12	13	15	16	17	19	20	21	22	24	25
YM.t	16	32	48	4	20	36	52	9	25	41	57	13	29	45	1	17	33	49	5	21

⌜ f. 2^{v} B

‖ ravikṣepakāḥ ‖

	1	2	3	4	5	6	7	8	9	10	11	12	13	14	15	⌜16	17	18	19	20	21	22	23	24	25	26	27	28	29	30
IM.z	7	4	0	9	5	2	10	7	3	0	8	5	2	10	7	3	0	8	5	1	10	6	3	0	8	5	1	10	6	3
IM.d	15	1	18	4	20	7	23	9	26	12	28	15	1	17	4	20	7	23	9	26	12	28	15	1	17	4	20	6	23	9
IM.m	33	53	14	34	55	15	35	56	16	37	57	18	38	58	19	39	0	20	41	1	21	42	2	23	43	3	24	44	5	25
IM.s	25	50	16	41	6	32	57	22	48	13	39	4	29	55	20	45	11	36	2	27	53	18	43	8	34	59	25	50	15	41
IM.t	22	44	7	29	52	14	37	59	21	44	6	29	51	13	36⌟c	58	21	43	6	28	50	13	35	58	20	43	5	27	50	12⌟d

⌟c f. 2^{r} B
⌟d f. 2^{v} B
f. 1^{r} Kh
f. 2^{r} S45
f. 1^{r} SMB

Table I: apparatus criticus

‖ ravidinābhogāḥ ‖] ‖ ravidinabhogaḥ ‖ (f. 2^r) B; rave- bhogyadināni Kh; ‖ ravidinābhogāḥ ‖ S45; ravibhogyadināni deśāntara SMB • ‖ ravimāsabhogāḥ ‖] ravimāsaṣepāḥ ‖ (f. 2^r) B; raver māsā Kh; ‖ ravimāsabhogāḥ ‖ S45; ‖ ravimāsa ‖ SMB • ‖ ravivarṣāṇi ‖] ravibaṣārṇaṣepā varṣaṣepāḥ ‖ (f. 2^r) B; raver varṣāṇi Kh; ‖ ravivarṣāṇi ‖ rāmabījakalā 2 dhanaṃ | ardhabhuktiḥ 29 | 34 ‖ SMB • ‖ ravikṣepakāḥ ‖] ravimadhyeṣepāḥ ‖ 10 | 29 | 13 | 0 | 22 (f. 2^r) B, ravikhepāsamāptā ‖ (f. 2^v) B; ravekṣepāḥ Kh; ‖ ravikṣepakā ‖ S45; ‖ ravikṣepakaḥ ‖ pratikoṣṭakakṣepaka 8 | 16 | 20 | 25 ‖ SMB.

DM.t(3) 31 Kh DM.s(4) 33 SMB; DM.t(4) 41 Kh DM.t(6) 2 Kh DM.t(7) 12 Kh DM.t(8) 23 Kh DM.t(9) 33 Kh DM.s(10) 20 Kh, 22 SMB; DM.t(10) 43 Kh DM.s(11) 28 Kh, 30 SMB; DM.t(11) 54 Kh DM.s(12) 48 Kh; DM.t(12) 14 Kh DM.s(13) 56 Kh; DM.t(13) 24 Kh DM.t(14) 30 Kh DM.t(15) 35 Kh DM.s(16) 11 SMB; DM.t(16) 45 Kh DM.s(17) 20 SMB; DM.t(17) 55 Kh DM.t(18) 6 Kh DM.t(19) 16 Kh DM.t(20) 27 Kh DM.t(21) 37 Kh DM.t(22) 12 Kh DM.s(23) 8 SMB DM.t(24) 5 B, Kh DM.s(28) 49 SMB; DM.t(28) 45 B DM.s(29) 57 SMB.

MM.t(2) 12 B MM.t(3) 18 B MM.m(4) 56 B; MM.t(4) 36 B MM.s(5) 21 Kh; MM.t(5) 25 Kh MM.t(6) 36 B MM.s(7) 36 SMB; MM.t(7) 42 B, 26 Kh MM.s(8) 32 Kh; 41 SMB; MM.t(8) 48 B, 57 Kh MM.s(9) 46 SMB; MM.t(9) 54 B MM.t(10) 0 B MM.t(11) 6 B MM.t(12) 12 B.

YM.t(1) 12 B YM.m(2) 48 Kh; YM.t(2) 24 B YM.t(3) 36 B YM.s(4) 4 B; YM.t(4) 48 B YM.t(5) 0 B YM.t(6) 12 B YM.s(7) 9 SMB; YM.t(7) 24 B YM.s(8) 8 Kh; YM.t(8) 36 B YM.s(9) 10 B; YM.t(9) 48 B YM.s(10) 13 SMB; YM.t(10) 0 B YM.s(11) 14 SMB; YM.t(11) 12 B YM.s(12) 14 B; YM.t(12) 24 B YM.s(13) 15 B; YM.t(13) 36 B; 30 Kh YM.s(14) 16 B, 18 SMB; YM.t(14) 48 B, 46 Kh YM.s(15) 18 B; YM.t(15) 0 B, 2 Kh YM.s(16) 19 B; YM.t(16) 12 B, 14 Kh YM.s(17) 20 B; YM.t(17) 24 B, 34 Kh YM.m(18) 41 Kh; YM.s(18) 21 B, 29 Kh, 23 SMB; YM.t(18) 36 B, 50 Kh YM.d(19) 27 Kh; YM.s(19) 22 B; YM.t(19) 48 B, 6 Kh YM.s(20) 22 B; YM.t(20) 0 B, 22 Kh.

IM.s(2) 51 SMB IM.s(4) 42 SMB IM.s(5) 7 SMB IM.t(7) 36 Kh IM.t(8) 58 Kh IM.s(10) 14 SMB; IM.t(10) 43 Kh IM.m(12) 28 Kh; IM.t(12) 28 Kh IM.s(13) 30 SMB IM.s(16) 46 SMB IM.s(18) 37 SMB IM.z(20) 11 SMB; IM.s(20) 29 SMB IM.s(21) 52 B, Kh; IM.t(21) 51 Kh IM.s(24) 34 Kh, 9 SMB; IM.t(24) 57 Kh IM.t(25) 4 Kh IM.t(26) 42 Kh IM.s(29) 16 SMB; IM.t(29) 42 Kh.

DM.#(0) to DM.#(22) *om.* B; "10 | 29 | 13 | 0 | 22" written by a different hand on f. 2^r B; IM-table has an inserted argument $\overset{31}{\wedge}$ written by a different hand on f. 2^v B; IM.#(31) $11^s\ 25^\circ\ 46'6''22'''$, IM.#(32) $8^s\ 12^\circ\ 6'31''44'''$, and an empty column for IM.#(33) written by a different hand in the left margin of f. 2^v B • DM.#(0) *om.* Kh; "ravermāsa" in the left margin of f. 1^r Kh • Table I *om.* S43 • "śloka 600" written by a different hand in the top margin of f. 2^r S45; DM.#(30) *om.* S45 • DM.#(0) *om.* SMB; DM.t(1–30), MM.t(1–12), YM.t(1–20), and IM.t(1–30) *om.* SMB; DM-table has additional entry | kṣe | 0 0 59 8 | in SMB; IM-table has an additional entry (corrective) $1^s\ 26^\circ\ 1'\ 29''$ for IM.#(20) in SMB.

Variant IM-table in S45:

30	31	32	33	34	35	36	37	38	39	40	41	42	43	44
3	11	8	4	1	10	6	3	11	8	4	1	9	6	2
9	25	12	28	14	1	17	3	20	6	22	9	25	11	28
25	46	6	26	47	7	28	48	9	29	49	10	30	51	11
41	6	31	57	22	48	13	38	4	19	54	20	45	11	36
12	34	57	19	41	4	26	48	11	33	55	18	40	2	25

45	46	47	48	49	50	51	52	53	54	55	56	57	58	59
11	8	4	1	9	6	2	11	7	4	0	9	6	2	11
14	0	17	3	19	6	22	8	25	11	20	14	0	16	3
32	52	12	33	53	14	34	54	15	35	55	16	37	57	17
1	27	52	17	43	8	34	59	24	50	15	40	6	31	57
47	9	32	54	16	39	1	23	46	8	30	53	15	37	0

Table II

⌜ f. 2v B
f. 1v Kh
f. 2v S45
f. 1v SMB

⌜‖ candradinabhogāḥ ‖ candrābdabījāni gatābdā 442 | 5 | 46 | 30 | o varṣa 78 kalā 1 vṛddhirṇam ‖

	1	2	3	4	5	6	7	8	9	10	11	12	13	14	15	16	17	18	19	20	21	22	23	24	25	26	27	28	29	30
DM.z	0	0	1	1	2	2	3	3	3	4	4	5	5	6	6	7	7	7	8	8	9	9	10	10	10	11	11	0	0	1
DM.d	13	26	9	22	5	19	2	15	28	11	24	8	21	4	17	0	13	27	10	23	6	19	3	16	29	12	25	8	22	5
DM.m	10	21	31	42	52	3	14	24	35	45	56	6	17	28	38	49	59	10	21	31	42	52	3	13	24	35	45	56	6	17
DM.s	34	9	44	19	54	29	4	39	13	48	23	58	33	8	43	18	52	27	2	37	12	47	22	57	31	6	41	16	51	26
DM.t	52	45	38	30	33	45	8	0	53	45	38	30	22	15	7	0	53	45	38	20	23	15	8	1	54	4	38	30	23	16⌟

⌟ f. 2v B

⌜ f. 3r B

⌜‖ candramāsāḥ‖

	1	2	3	4	5	6	7	8	9	10	11	12
MM.z	1	2	3	4	5	7	8	9	10	11	0	2
MM.d	5	10	15	21	26	1	7	12	17	22	28	3
MM.m	17	34	52	9	27	44	2	19	36	54	11	29
MM.s	26	52	18	45	11	37	3	30	56	22	48	15
MM.t	16	32	48	4	19	35	15	7	23	39	55	11

‖ candravarṣāṇi ‖

	1	2	3	4	5	6	7	8	9	10	11	12	13	14	15	16	17	18	19	20
YM.z	2	4	6	8	10	0	2	4	7	9	11	1	3	5	7	9	11	2	4	6
YM.d	3	6	10	13	17	20	24	27	1	4	8	11	15	18	22	25	29	2	6	9
YM.m	29	58	27	57	26	55	24	54	23	52	21	51	20	49	18	48	17	46	15	45
YM.s	15	30	45	0	15	31	46	1	16	31	47	2	17	32	47	2	18	33	48	3
YM.t	11	22	33	44	55	6	17	28	39	50	1	12	20	35	46	57	8	19	30	41

⌜ f. 3v B

‖ candrakṣepakāḥ ‖

	1	2	3	4	5	6	7	8	9	⌜10	11	12	13	14	15	16	17	18	19	20	21	22	23	24	25	26	27	28	29	30
IM.z	5	11	5	0	6	0	7	1	7	2	8	2	9	3	9	4	10	4	11	5	11	6	0	6	1	7	1	8	2	8
IM.d	8	18	28	8	17	27	7	17	26	6	16	26	5	15	25	5	14	24	4	14	23	3	13	23	2	12	22	2	11	21
IM.m	50	35	21	6	51	36	21	6	51	36	21	6	21	36	21	6	51	36	22	7	52	37	22	7	52	37	22	7	52	37
IM.s	53	57	1	4	8	12	15	14	23	26	30	34	38	41	45	49	52	56	0	3	7	11	14	18	22	26	29	33	37	41
IM.t	41	23	4	46	58	9	51	32	14⌟a	56	37	19	0	42	42	5	52	34	16	18	40	22	54	46	28	10	43	34	16	22⌟b

⌟a f. 3r B
⌟b f. 3v B
f. 1v Kh
f. 2v S45
f. 1v SMB

Table II: apparatus criticus

‖ candradinābhogā ‖] ‖ atha candradīnabhogāḥ ‖ (f. 2ᵛ) B, iti candrabhogāsamāḥtā ‖ (f. 2ᵛ) B; vidhor bhogyadināni Kh; candrabhogyadināni deśāntara 296 candravikalā madhye dhanaṃ ‖ ardhabhuktiḥ ‖ SMB • ‖ candramāsāḥ‖] candramāsā ‖ (f. 3ʳ) B; candrasya māsāḥ Kh • ‖ candravarṣāṇi ‖] candrabarṣā (f. 3ʳ) B; vidhor varṣāṇi Kh; ‖ candravarṣāṇi | rāmabījaṃkalā 15 ṛṇaṃ ‖ SMB • ‖ candrakṣepakāḥ ‖] candraṣepāḥ ‖ 10 | 29 | 5 | 50 ‖ (f. 3ʳ) B, candraṣepāsamāptā ‖ (f. 3ᵛ) B; ‖ vidhokṣepāḥ madhyamāḥ ‖ Kh.

DM.s(1) 35 SMB DM.s(2) 10 SMB DM.z(4) 1 B DM.t(5) 23 B, Kh DM.t(6) 15 B, Kh DM.s(8) 29 B DM.s(9) 14 SMB; DM.t(9) 52 B, Kh DM.d(10) 8 SMB; DM.m(10) 6 SMB; DM.s(10) 58 SMB DM.s(11) 28 B DM.m(12) 6 B DM.s(13) 17 B, 23 SMB DM.s(16) 19 B DM.s(17) 53 SMB DM.s(18) 28 SMB DM.s(19) 20 S45 DM.t(20) 30 B, 37 Kh DM.s(21) 13 SMB, Kh DM.t(24) 6 Kh DM.z(25) 11 Kh; DM.s(25) 21 Kh, 32 SMB; DM.t(25) 53 B, Kh DM.d(26) 22 B; DM.s(26) 7 SMB; DM.t(26) 45 B, Kh DM.d(28) [-] Kh DM.m(29) 16 Kh DM.t(30) 15 B, Kh.

MM.t(2) 31 B MM.t(3) 47 B, Kh MM.d(3) 19 SMB MM.d(4) 17 S45; MM.t(4) 3 B, Kh MM.d(6) 0 B MM.s(7) 4 SMB; MM.t(7) 51 B, Kh MM.m(11) 22 SMB; MM.s(11) 4 SMB.

YM.s(4) 1 SMB YM.s(5) 16 SMB YM.m(7) 34 S45; YM.s(7) 43 B YM.s(10) 32 SMB YM.m(12) 52 SMB YM.z(13) 2 Kh; YM.s(13) 27 B; YM.t(13) 23 B, Kh YM.d(14) 14 S45; YM.m(14) 41 Kh; YM.s(14) 31 Kh YM.s(15) 48 B, SMB; YM.t(15) $\frac{2}{57}$ Kh YM.d(16) 45 Kh; YM.s(16) 3 SMB YM.t(17) 7 B YM.s(20) 4 SMB; YM.t(20) 45 Kh.

IM.s(1) 54 SMB IM.z(4) 1 Kh IM.d(5) 13 Kh; IM.t(5) 28 Kh IM.z(7) 1 B, 0 Kh; IM.d(7) 17 B; IM.s(7) 6 B; IM.t(7) 32 B IM.s(8) 19 B, 19 Kh, SMB IM.s(9) 30 SMB IM.s(10) 27 SMB; IM.t(10) 46 Kh IM.m(12) 4 Kh; IM.s(12) 29 SMB IM.m(13) 51 Kh, SMB IM.z(14) 4 Kh; IM.s(14) 42 SMB IM.d(15) 15 B; IM.m(15) 36 B; IM.s(15) 41 B; IM.t(15) 24 Kh IM.d(16) 4 Kh IM.s(17) 53 SMB; IM.t(17) 47 Kh IM.t(18) 38 Kh IM.t(19) 10 Kh IM.s(20) 4 SMB; IM.t(20) 52 Kh IM.t(21) 33 Kh IM.t(22) 12 Kh IM.s(23) 15 SMB; IM.t(23) 56 Kh IM.t(24) 38 Kh IM.t(25) 30 Kh IM.t(26) 1 Kh IM.s(27) 30 SMB IM.t(28) 24 Kh IM.s(29) 38 Kh; IM.t(29) 7 Kh IM.m(30) 30 B; IM.s(30) 40 Kh; IM.t(30) 46 Kh.

IM-table has an empty column for IM.#(31) on f. 3ᵛ B; "10 | 29 | 5 | 50" written by a different hand on f. 3ʳ B; IM.#(32) $8^s\ 11^\circ\ 7'\ 47''\ 52'''$ written by a different hand in the right margin of f. 3ᵛ B • Table II *om.* S43 • DM.#(30) *om.* S45 • DM.t(1–30), MM.t(1–12), YM.t(1–20), and IM.t(1–30) *om.* SMB; IM-table has additional entry | kṣe | 6 9 44 0 | in SMB.

Variant IM-table in S45:

30	31	32	33	34	35	36	37	38	39	40	41	42	43	44	45	46	47	48	49	50	51	52	53	54	55	56	57	58	59	60
8	3	9	3	10	4	10	4	11	5	11	6	0	6	1	7	1	8	2	8	3	9	3	10	4	10	5	11	5	0	6
21	1	11	20	0	10	20	29	9	19	29	8	18	28	8	17	27	7	17	26	6	16	26	5	15	25	5	14	24	4	14
37	22	7	52	37	22	8	53	38	23	8	53	38	23	8	53	38	23	8	53	38	23	9	54	39	24	9	54	39	24	9
40	44	48	51	55	59	2	6	10	14	17	21	25	28	32	36	39	43	47	50	54	58	2	5	9	13	16	20	24	27	31
48	30	12	53	35	16	58	40	21	3	44	26	8	49	31	12	54	36	17	59	40	22	4	45	27	8	50	31	13	55	36

Table III

⌜a f. 3^{v} B, f. 1^{v} Kh, f. 3^{r} S45, f. 2^{r} SMB
⌜b f. 2^{r} Kh
⌜c f. 4^{r} B

⌜a‖ candroccadināni ‖ uccābdabījānigatābda 442 | 14 | 45 | 15 gatavarṣa 30 sadā dhanaṃ ‖

	0	1	2	3	4	5	6	7	8	9	10	11	⌜b12	13	14	15	16	17	18	19	20	21	22	⌜c23	24	25	26	27	28	29	30
DM.z	0	0	0	0	0	0	0	0	0	0	0	0	0	0	0	0	0	0	0	0	0	0	0	0	0	0	0	0	0	0	0
DM.d	0	0	0	0	0	0	0	0	0	1	1	1	1	1	1	1	1	1	2	2	2	2	2	2	2	2	2	3	3	3	3
DM.m	0	6	13	20	26	33	40	46	53	0	6	13	20	26	33	40	46	53	0	6	13	20	26	33	40	47	53	0	7	13	20
DM.s	0	40	21	2	43	24	5	46	27	8	48	29	10	51	32	13	54	35	16	57	37	18	59	40	21	2	43	24	5	46	26
DM.t	0	53	46	41	53	29	23	16	10	4	52	52⌟d	46	40	33	27	21	15	9	2	56	50	48⌟e	38	32	26	19	13	7	1	55

⌟d f. 1^{v} Kh
⌟e f. 3^{v} B

‖ candroccamāsāḥ‖

	1	2	3	4	5	6	7	8	9	10	11	12
MM.z	0	0	0	0	0	0	0	0	1	1	1	1
MM.d	3	6	10	13	16	20	23	26	0	3	6	10
MM.m	20	40	1	21	42	2	23	43	4	24	44	5
MM.s	26	53	20	47	14	41	8	35	2	29	56	23
MM.t	55	50	45	40	35	31	26	21	16	11	6	1

‖ candroccavarṣābhogāḥ ‖

	1	2	3	4	5	6	7	8	9	10	11	12	13	14	15	16	17	18	19	20
YM.z	1	2	4	5	6	8	9	10	0	1	2	4	5	6	8	9	10	0	1	2
YM.d	10	20	0	10	20	0	10	20	0	10	20	1	11	21	1	11	21	1	11	21
YM.m	5	10	16	21	26	32	37	43	48	53	59	4	9	15	20	26	31	36	42	47
YM.s	23	46	9	32	55	18	41	4	27	50	13	36	59	22	45	8	31	54	17	40
YM.t	1	3	5	7	9	11	12	14	16	18	20	22	23	25	27	29	31	33	35	36

⌜ f. 4^{v} B

‖ candroccakṣepakāḥ ‖

	1	2	3	4	5	6	7	8	9	10	11	12	13	14	15	16	17	18	19	20	21	⌜22	23	24	25	26	27	28	29	30
IM.z	7	9	0	3	6	8	11	2	5	7	10	1	3	6	9	0	2	5	8	11	1	4	7	9	0	3	6	8	11	2
IM.d	7	28	20	12	4	25	17	9	1	23	14	6	28	20	12	3	24	17	9	1	22	14	6	28	20	11	3	25	17	9
IM.m	0	48	36	23	11	59	46	34	22	9	57	45	32	20	8	55	43	31	18	6	54	41	29	17	4	52	40	27	15	3
IM.s	39	20	0	41	22	3	43	24	5	45	26	7	47	28	9	50	29	11	52	32	13	54	34	15	54	35	27	58	39	17
IM.t	31	13	50	27	30	41	43	25	37	49	31	13	55	37	19	1	43	25	7	49	31⌟f	13	55	37	19	51	43	25	7	49⌟g

⌟f f. 4^{r} B
⌟g f. 4^{v} B, f. 2^{r} Kh, f. 3^{r} S45, f. 2^{r} SMB

Table III: apparatus criticus

|| candroccadināni ||°…°dhanaṃ] || uccabhogāḥ || (f. 3^{v}) B, uccabhogasamāptāḥ || (f. 4^{r}) B; candroccamadhyadinabhogāḥ || (f. 2^{r}) Kh; candrocca-bhogyadināni || ardhabhuktiḥ 3 | 21 || deśāntara 2 vikalā dhanaṃ || SMB • || candroccamāsāḥ||] uccamāśā (f. 4^{r}) B; candroccasya māsāḥ || (f. 2^{r}) Kh; || candroccamāsā || SMB • || candroccavarṣābhogāḥ ||] uccavarṣāṇiḥ || ṣepa[-]haliḥ || varṣāṇi samāptāḥ || (f. 4^{v}) B; candroccavarṣāṇi || (f. 2^{r}) Kh; || candroccavarṣāṇi || rāmabījakalāḥ 30 dhanaṃ || SMB • || candroccakṣepakāḥ ||] uccakhepāḥ 4 | 15 | 12 | 59 (f. 4^{r}) B, uccekhepā samāpatāḥ || (f. 4^{v}) B; candroccakṣepakādi (f. 2^{r}) Kh.

DM.s(1) 41 SMB DM.s(2) 23 SMB; DM.t(2) 47 B, Kh DM.s(3) 3 SMB DM.t(4) 35 B, Kh DM.m(5) 23 SMB; DM.t(5) 25 B DM.t(6) 18 B DM.t(7) 11 B DM.t(8) 4 B DM.t(9) 7 B DM.s(10) 49 B, SMB; DM.t(10) 48 B, 56 Kh DM.s(11) 30 B, SMB; DM.t(11) 43 B, 49 Kh DM.s(12) 11 B, SMB; DM.t(12) 36 B, 42 Kh DM.s(13) 52 B, SMB; DM.t(13) 36 B, 35 Kh DM.s(14) 33 B; DM.t(14) 22 B, 28 Kh DM.s(15) 14 B; DM.t(15) 15 B DM.s(16) 58 B; DM.t(16) 28 B, 24 Kh DM.t(17) 1 B DM.s(18) 15 B; DM.t(18) 54 B, 8 Kh DM.s(19) 56 B; DM.t(19) 47 B DM.s(20) 30 SMB; DM.t(20) 40 B DM.s(21) 19 SMB; DM.t(21) 53 B DM.s(22) 49 SMB; DM.t(22) 26 B, 44 Kh DM.t(23) 39 B DM.s(24) 31 SMB; DM.t(24) 12 B DM.t(25) 5 B DM.t(26) 58 B DM.s(27) 23 B, 14 Kh; DM.t(27) 53 B DM.s(28) 4 B; DM.t(28) 4 B DM.s(29) 45 B; DM.t(29) 37 B, 7 Kh DM.s(30) 27 SMB; DM.t(30) 30 B.

MM.s(1) 27 SMB MM.s(2) 54 SMB MM.s(3) 21 SMB MM.s(4) 48 SMB MM.s(6) 21 B MM.t(7) 24 Kh MM.m(8) 42 Kh MM.s(10) 2^{1}_{1} Kh MM.t(12) 0 B.

YM.s(1) 33 B YM.t(2) 2 Kh YM.s(3) 8 Kh YM.t(5) 8 Kh YM.m(8) 46 SMB YM.m(13) 8 Kh; YM.s(13) 58 B YM.t(17) 36 Kh YM.t(18) 13 Kh YM.t(19) 36 B, Kh YM.t(20) 27 Kh.

IM.t(1) 37 Kh IM.t(2) 24 Kh IM.s(3) 1 SMB; IM.t(3) 51 Kh IM.t(4) 28 Kh IM.t(5) 3 B, 5 Kh IM.s(6) 2 Kh; IM.t(6) 42 Kh IM.t(7) 19 Kh IM.s(8) 23 Kh; IM.s(8) 56 Kh IM.s(9) 4 Kh, SMB; IM.t(9) 33 Kh IM.t(10) 10 Kh IM.s(11) 25 Kh; IM.t(11) 47 Kh IM.z(12) 2 Kh; IM.s(12) 6 Kh, SMB; IM.t(12) 25 Kh IM.t(13) 2 Kh IM.s(14) 27 Kh, SMB; IM.t(14) 39 Kh IM.s(15) 8 Kh, SMB; IM.t(15) 16 Kh IM.s(16) 48 Kh, 49 SMB; IM.t(16) 53 Kh IM.d(17) 25 Kh, SMB; IM.s(17) 30 B; IM.t(17) 30 Kh IM.s(18) 10 Kh, SMB; IM.t(18) 7 Kh IM.s(19) 50 Kh, 51 SMB; IM.t(19) 44 Kh IM.s(20) 31 Kh, SMB; IM.t(20) 21 Kh IM.s(21) 11 Kh, 12 SMB; IM.t(21) 58 Kh IM.s(22) 54 Kh, SMB; IM.t(22) 35 Kh IM.s(23) 33 Kh, SMB; IM.t(23) 14 Kh IM.s(24) 13 Kh, 14 SMB; IM.t(24) 50 Kh IM.s(25) 16 B; IM.t(25) 27 Kh IM.s(26) 37 B; IM.t(26) 4 Kh IM.s(27) 15 Kh, 16 SMB; IM.t(27) 31 Kh IM.s(28) 57 Kh, SMB; IM.t(28) 10 Kh IM.s(29) 36 Kh, 37 SMB; IM.t(29) 45 Kh IM.m(30) 13 Kh; IM.t(30) 32 Kh.

DM.#(0) *om.* B; the table YM.#(1–20) is transposed and placed after the IM-table on f. 4^{v} B; "4 | 15 | 12 | 59" written by a different hand on f. 4^{r} B • DM.#(0) *om.* Kh; • Table III *om.* S43 • DM.#(30) *om.* S45 • DM.#(0) *om.* SMB; DM.t(1–30), MM.t(1–12), YM.t(1–20), and IM.t(1–30) *om.* SMB; DM.#(31), DM.#(32), and DM.#(33) are empty columns in the DM-table of SMB.

Variant IM-table in S45:

30	31	32	33	34	35	36	37	38	39	40	41	42	43	44	45
2	5	7	10	1	3	6	9	0	2	5	8	11	1	4	7
9	0	22	14	6	28	19	11	3	25	17	8	0	22	14	5
3	50	38	26	13	1	49	37	24	12	0	47	35	23	10	58
17	57	38	19	59	40	21	1	42	22	3	44	24	5	45	26
22	59	36	13	50	26	3	40	17	54	30	7	44	21	48	34

46	47	48	49	50	51	52	53	54	55	56	57	58	59	60
9	0	3	6	8	11	2	5	7	10	1	3	6	9	0
27	19	11	3	24	16	8	0	22	13	5	27	19	11	2
46	33	21	9	56	44	32	19	7	55	42	30	18	5	53
7	47	28	9	49	30	10	51	32	12	53	33	14	55	35
11	48	25	2	38	15	52	29	5	42	19	56	33	10	46

Table IV

⌜a f. 4v B, f. 2r Kh, f. 3v S45, f. 2v SMB; ⌜b f. 5r B; ⌜c f. 2v Kh; ⌟d f. 4v B; ⌟e f. 2r Kh

⌜a‖ pātadinabhogāḥ ‖

	0	1	2	3	4	5	6	7	8	9	10	11	12	⌜b13	14	15	16	17	18	19	20	⌜c21	22	23	24	25	26	27	28	29	30
DM.z	0	0	0	0	0	0	0	0	0	0	0	0	0	0	0	0	0	0	0	0	0	0	0	0	0	0	0	0	0	0	0
DM.d	0	0	0	0	0	0	0	0	0	0	0	0	0	0	0	0	0	0	0	1	1	1	1	1	1	1	1	1	1	1	1
DM.m	0	3	6	9	12	15	19	22	25	28	31	34	38	41	44	47	50	54	57	0	3	6	9	13	16	19	22	25	29	32	35
DM.s	0	10	21	32	43	54	4	15	26	37	48	58	9	20	31	42	52	3	14	25	36	46	57	8	19	30	40	51	2	13	24
DM.t	0	48	36	25	13	2	50	39	27	15	4	52	42⌟d	29	18	6	54	43	31	20	8⌟d	57	45	33	22	10	59	47	36	24	12

⌜ f. 5v B

‖ pātamāsāḥ ‖

	1	2	3	4	5	6	7	8	9	10	11	12
MM.z	0	0	0	0	0	0	0	0	0	0	0	0
MM.d	1	3	4	6	7	9	11	12	14	15	17	19
MM.m	35	10	46	21	57	32	7	43	18	54	29	4
MM.s	24	48	12	36	1	25	49	13	37	2	26	50
MM.t	12	25	36	50	3	16	28	41	53	6	19	33

‖ pātavarṣabhogāḥ ‖

	1	2	3	4	5	6	7	8	9	10	11	12	13	14	15	⌜16	17	18	19	20
YM.z	0	1	1	2	3	3	4	5	5	6	6	7	8	8	9	10	10	11	0	0
YM.d	19	8	27	16	5	24	13	2	21	10	29	18	8	27	16	5	24	13	2	21
YM.m	4	9	14	19	24	29	33	38	43	48	53	58	2	7	12	17	22	27	32	36
YM.s	50	41	31	22	12	3	53	44	34	25	15	6	56	47	37	28	18	9	0	50
YM.t	33	3	34	6	48	9	41	12	44	15	47	18	57	22	53⌟	25	56	28	0	31

⌟ f. 5r B

‖ pātakṣepakāḥ ‖

	1	2	3	4	5	6	7	8	9	10	11	12	13	14	15	16	17	18	19	20	21	22	23	24	25	26	27	28	29	30
IM.z	10	11	11	0	1	1	2	3	4	4	5	6	6	7	8	9	9	10	11	11	0	1	2	2	3	4	5	5	6	7
IM.d	9	0	22	13	5	27	18	10	1	23	15	6	28	20	11	3	24	16	8	29	21	12	4	26	17	9	0	22	14	5
IM.m	1	38	15	52	29	6	43	19	56	33	10	47	24	0	37	14	51	28	5	41	18	55	32	9	46	23	59	36	13	50
IM.s	59	50	40	21	31	12	3	53	43	34	24	15	5	56	46	37	28	18	9	59	50	40	31	21	12	2	53	46	34	24
IM.t	31	3	35	35	38	9	43	12	44	16	51	19	50	22	54	25	31	28	7	39	11	43	15	49	19	53	23	55	26	47⌟

⌟ f. 5v B, f. 2v Kh, f. 3v S45, f. 2v SMB

Table IV: apparatus criticus

‖ pātadinabhogāḥ ‖] ‖ pātabhaugāḥ ‖ (f. 4ᵛ) B, pātabhaugasamāptāḥ ‖ (f. 5ʳ) B; pātasya bhogyadināni (f. 2ʳ) Kh; pātabhogyadināni ‖ ardhabhukti 1 | 36 ‖ deśāntaravikalā 1 dhanaṃ ‖ SMB • ‖ pātamāsāḥ ‖] pātamāśāḥ māśāpūrṇaḥ (f. 5ʳ) B • ‖ pātavarṣabhogāḥ ‖] pātavarṣāṇīḥ ‖ (f. 5ʳ) B; [-] ‖ pātasya varṣāṇi (f. 2ᵛ) Kh; ‖ pātavarṣāṇi ‖ rāmabījaṃ 30 kalā ṛṇaṃ ‖ SMB • ‖ pātakṣepakāḥ ‖] pātakhepāḥ 9 | 17 | 25 | 9 pātaṣepāsaṃpūrṇaḥ ‖ (f. 5ᵛ) B; ‖ atha pātamandakṣepāḥ ‖ (f. 2ᵛ) Kh.

DM.s(1) 11 SMB DM.t(3) 24 B DM.t(4) 12 B DM.t(5) 0 B DM.s(6) 5 SMB; DM.t(6) 48 B DM.t(7) 36 B, 38 Kh DM.s(8) 36 B; DM.t(8) 24 B, 30 Kh DM.t(9) 12 B DM.t(10) 0 B DM.s(11) 59 SMB; DM.t(11) 48 B DM.s(12) 10 SMB; DM.t(12) 36 B, 41 Kh DM.t(13) 24 B DM.m(14) 45 B; DM.t(14) 12 B, 17 Kh DM.t(15) 0 B, 9 Kh DM.m(16) 53 SMB; DM.t(16) 48 B DM.s(17) 4 SMB; DM.t(17) 36 B DM.s(18) 34 Kh; DM.t(18) 24 B DM.t(19) 12 B DM.t(20) 0 B DM.s(21) 56 B, 47 SMB; DM.t(21) 48 B, 56 Kh DM.s(22) 58 SMB; DM.t(22) 36 B, 47 Kh DM.t(23) 24 B DM.t(24) 12 B, 32 Kh DM.t(25) 0 B DM.s(26) 4- SMB; DM.t(26) 49 B, 58 Kh DM.s(27) 52 SMB; DM.t(27) 36 B DM.t(28) 24 B, 35 Kh DM.m(29) 33 B; DM.t(29) 12 B DM.t(30) 0 B.

MM.t(2) 24 Kh MM.t(3) 37 Kh MM.d(4) 1 Kh; MM.s(4) 37 SMB; MM.t(4) 48 B MM.s(5) 2 S45; MM.t(5) 0 B MM.m(6) 33 Kh; MM.t(6) 12 B, 15 Kh MM.t(7) 24 B MM.s(8) 14 SMB; MM.t(8) 36 B MM.s(9) 38 SMB; MM.t(9) 46 B, 56 Kh MM.t(10) 0 B MM.t(11) 12 B, 18 Kh MM.t(12) 24 B, 31 Kh.

YM.t(1) 31 B, Kh YM.s(3) 21 SMB YM.t(4) 4 B YM.t(5) 38 B, Kh YM.m(6) 19 Kh YM.s(7) 54 SMB YM.m(9) 41 Kh; YM.s(9) 35 SMB YM.s(11) 13 SMB YM.s(13) 57 SMB; YM.t(13) 50 B, Kh YM.m(16) 22 B; YM.t(16) 56 B YM.s(17) 19 SMB; YM.t(17) 51 Kh YM.m(19) 31 B; YM.s(19) 59 B, Kh; YM.t(19) 49 B, 51 Kh.

IM.m(2) 39 B IM.s(4) 40 B; IM.t(4) 8 Kh IM.s(5) 31 B; IM.t(5) 39 Kh IM.s(6) 21 B; IM.t(6) 40 Kh IM.s(7) 12 B, 2 Kh; IM.t(7) 41 Kh IM.s(9) 42 B, 44 SMB IM.s(10) 36 Kh IM.t(11) 47 Kh IM.s(11) 25 SMB IM.s(13) 6 SMB IM.m(15) 57 Kh; IM.s(15) 47 SMB IM.m(16) 54 SMB IM.s(17) 27 Kh; IM.t(17) 57 Kh IM.z(19) 10 SMB; IM.t(19) 0 Kh IM.t(20) 31 Kh IM.m(21) 19 Kh; IM.t(21) 3 Kh IM.t(22) 35 Kh IM.m(23) 34 B; IM.t(23) 6 Kh IM.t(24) 38 Kh IM.m(25) 48 Kh; IM.s(25) 9 Kh IM.s(26) 3 SMB; IM.t(26) 41 Kh IM.t(27) 12 Kh IM.s(28) 43 Kh, 44 SMB; IM.t(28) 44 Kh IM.t(29) 16 Kh IM.s(30) 25 SMB.

DM.#(0) *om.* B; IM-table has the correction "9 | 17 | 25 | 9" for "7 | 21 | 24 | 21", both written by a different hand on f. 5ᵛ B; IM-table has the correction $0^s\ 21^\circ\ 18'\ 50''\ 11'''$ for IM.#(21), written below the tabulated entry in the bottom margin of f. 5ᵛ B • DM.#(0) *om.* Kh • Table IV *om.* S43 • DM.#(30) *om.* S45 • DM.#(0) *om.* SMB; DM.t(1–30), MM.t(1–12), YM.t(1–20), and IM.t(1–30) *om.* SMB.

Variant IM-table in S45:

30	31	32	33	34	35	36	37	38	39	40	41	42	43	44	45	46	47	48	49	50	51	52	53	54	55	56	57	58	59	60
7	7	8	9	10	10	11	0	0	1	2	3	3	4	5	6	6	7	8	8	9	10	11	11	0	1	1	2	3	4	4
5	27	19	10	2	23	15	7	28	20	11	3	25	16	8	0	21	13	4	26	18	9	1	22	14	6	27	19	11	2	24
50	27	4	40	17	54	31	8	45	21	58	35	12	49	26	3	39	16	53	30	7	44	20	57	34	11	48	25	1	38	15
24	15	5	56	46	37	27	18	9	59	50	40	31	21	12	2	53	43	34	24	15	5	56	46	37	27	18	9	59	50	40
47	19	51	22	54	26	57	29	1	32	3	35	7	38	10	42	13	45	16	47	19	50	22	54	25	57	29	0	32	4	35

Table v

⌜a f. 5v B, f. 2v Kh, f. 4r S45, f. 3r SMB
⌜b f. 6r B

⌟c f. 5v B
⌟d f. 2v Kh

⌜a‖ bhaumadinabhogāḥ ‖

	0	1	2	3	4	5	6	7	8	9	10	⌜b11	12	13	14	15	16	17	18	19	20	21	22	23	24	25	26	27	28	29	30
DM.z	0	0	0	0	0	0	0	0	0	0	0	0	0	0	0	0	0	0	0	0	0	0	0	0	0	0	0	0	0	0	0
DM.d	0	0	1	1	2	2	3	3	4	4	5	5	6	6	7	7	8	8	9	9	10	11	11	12	12	13	13	14	14	15	15
DM.m	0	31	2	34	5	37	8	40	11	42	14	45	17	48	20	51	23	54	25	57	28	0	31	3	34	6	37	8	40	11	43
DM.s	0	26	52	19	45	12	38	5	31	58	24	51	17	44	10	37	4	29	56	22	49	15	42	8	35	1	28	54	21	47	14
DM.t	0	28	56	24	42	22	49	17	45	13	41⌟c	14	38	6	34	2	30	59	27	55	23	15	20	48	16	44	12	40	8	37	4⌟d

⌜e f. 3r Kh
⌜f f. 6v B

⌟ f. 6r B

‖ bhaumamāsabhogāḥ ‖

	⌜e1	2	3	4	5	6	7	8	9	10	11	12
MM.z	0	1	1	2	2	3	3	4	4	5	5	6
MM.d	15	1	17	2	18	4	20	5	21	7	22	8
MM.m	43	26	9	52	36	19	2	45	29	12	55	38
MM.s	14	28	42	56	10	24	38	52	6	20	34	48
MM.t	4	10	15	20	24	29	34	39	44	49	34	58

‖ bhaumavarṣabhogāḥ ‖

	1	2	3	4	5	6	7	8	9	10	11	12	⌜f13	14	15	16	17	18	19	20
YM.z	6	0	6	1	7	1	8	2	8	2	9	3	9	4	10	4	10	5	11	5
YM.d	8	17	25	4	13	21	0	9	17	26	5	13	22	1	9	18	26	5	14	22
YM.m	38	17	56	35	14	52	31	10	49	28	6	45	24	3	42	21	59	38	17	56
YM.s	48	37	26	15	4	53	42	31	20	9	58	47	36	25	14	3	52	41	30	19
YM.t	58	57	55	54	53	51	50	48	47	45	44	42⌟	41	41	38	37	35	34	32	30

⌟ f. 6v B, f. 3r Kh, f. 4r S45, f. 3r SMB

‖ bhaumakṣepakāḥ ‖

	1	2	3	4	5	6	7	8	9	10	11	12	13	14	15	16	17	18	19	20	21	22	23	24	25	26	27	28	29	30
IM.z	1	7	1	6	0	6	0	5	11	5	11	4	10	4	10	3	9	3	9	3	8	2	8	2	7	1	7	1	6	0
IM.d	14	7	0	23	16	9	1	24	17	10	3	26	19	12	5	28	21	14	7	0	23	16	8	1	24	17	10	3	26	19
IM.m	20	17	13	9	5	2	58	54	51	47	43	40	36	33	29	25	21	18	14	10	7	3	59	56	52	48	45	41	37	34
IM.s	40	0	19	39	58	18	37	46	16	35	55	15	34	54	13	33	52	41	31	51	10	30	49	9	28	48	7	27	47	6
IM.t	31	2	33	3	34	5	36	57	37	59	30	1	31	3	24	55	26	57	28	59	30	1	32	3	34	15	36	7	38	9⌟

Table v: apparatus criticus

‖ bhaumadinabhogāḥ ‖] ‖ bhaumabhogāḥ ‖ (f. 5ᵛ) B, bhogapūrṇaḥ ‖ (f. 6ʳ) B; atha bhaumasya madhyabhogyadināni (f. 2ᵛ) Kh; bhaumabhogyadināni ‖ ardhabhuktiḥ 15 | 43 ‖ deśāntara 11 vikalā dhanaṃ ‖ SMB • ‖ bhaumamāsabhogāḥ ‖] bhaumamāśāḥ (f. 6ʳ) B; bhaumasya madhyamamāsāḥ (f. 3ʳ) Kh; ‖ bhaumamāsāḥ ‖ SMB • ‖ bhaumavarṣabhogāḥ ‖] varṣāṇiḥ ‖ (f. 6ʳ) B; ‖ bhaumavarṣāṇi ‖ rāmabījaṃ 80 kalādhanaṃ ‖ SMB • ‖ bhaumakṣepakāḥ ‖] bhaumakhepāḥ 7 | 21 | 24 | 29 (f. 6ᵛ) B; atha bhaumasya kṣepakāḥ (f. 3ʳ) Kh.

DM.1(t) 18 Kh DM.s(2) 53 SMB DM.s(3) 18 B; DM.t(3) 26 Kh DM.t(4) 52 B, Kh DM.t(5) 21 B, 21 Kh DM.s(6) 39 SMB; DM.t(6) 41 Kh DM.s(7) 45 B DM.s(8) 32 SMB DM.s(10) 25 SMB DM.t(11) 10 B, Kh DM.d(12) 5 SMB DM.s(13) 46 Kh DM.t(15) 0 B, 25 Kh DM.s(16) 3 B; DM.t(16) 28 B, 20 Kh DM.m(17) 50 SMB; DM.s(17) 30 SMB DM.t(18) 24 B DM.s(19) 23 SMB; DM.t(19) 52 B DM.t(20) 20 B DM.s(21) 16 SMB; DM.t(21) 48 B, 51 Kh DM.t(22) 16 B DM.s(23) 9 SMB; DM.t(23) 44 B DM.t(24) 12 B DM.s(25) 2 SMB; DM.t(25) 40 B DM.t(26) 8 B DM.s(27) 55 SMB; DM.t(27) 36 B DM.s(28) 1 Kh; DM.t(28) 4 B DM.t(29) 32 B DM.t(30) 0 B, 5 Kh.

MM.t(1) 5 B, Kh MM.s(2) 20 B MM.t(6) 19 B MM.z(7) 2 Kh MM.d(9) 20 Kh; MM.m(9) 22 Kh; MM.s(9) 7 SMB MM.s(10) 21 B, SMB MM.s(11) 35 SMB; MM.t(11) 54 B, 14 Kh MM.s(12) 49 SMB.

YM.z(1) 0 Kh; YM.d(1) 17 Kh; YM.m(1) 17 Kh; YM.s(1) 37 Kh, 49 SMB; YM.t(1) 57 Kh YM.s(2) 38 SMB; YM.t(2) 57 Kh YM.s(3) 27 SMB YM.s(4) 16 SMB YM.s(5) 5 SMB; YM.t(5) 57 Kh YM.s(6) 54 SMB YM.s(7) 43 SMB; YM.t(7) 55 B YM.s(8) 32 SMB YM.z(9) 9 Kh; YM.s(9) 21 SMB YM.s(10) 10 SMB YM.s(11) 59 SMB YM.s(12) 48 SMB; YM.t(12) 52 B YM.s(13) 37 SMB YM.s(14) 26 SMB; YM.t(14) 40 B, Kh YM.t(16) 36 Kh YM.t(17) 38 B YM.d(20) 20 B; YM.t(20) 31 B.

IM.s(1) 50 B IM.t(3) 32 Kh IM.s(4) 29 Kh, SMB IM.s(8) 17 Kh, 57 SMB; IM.t(8) 17 Kh IM.d(10) 11 Kh; IM.s(10) 36 Kh, SMB; IM.t(10) 8 Kh IM.d(11) 2 SMB; IM.t(11) 39 Kh IM.t(12) 10 Kh IM.m(13) 37 B; IM.t(13) 41 Kh IM.m(14) 32 Kh, SMB; IM.t(14) 11 Kh IM.s(15) 14 SMB; IM.t(15) 52 Kh IM.d(16) 18 B; IM.t(16) 23 Kh IM.s(17) 53 SMB; IM.t(17) 44 Kh IM.s(18) 12 Kh, SMB; IM.t(18) 14 Kh IM.t(19) 45 Kh IM.t(20) 16 Kh IM.d(21) 29 B; IM.s(21) 51 SMB; IM.t(21) 47 Kh IM.t(22) 18 Kh IM.s(23) 50 SMB; IM.t(23) 48 Kh IM.t(24) 18 Kh IM.s(25) 29 SMB; IM.t(25) 50 Kh IM.m(26) 49 Kh, 47 SMB; IM.t(26) 21 Kh IM.t(27) 52 Kh IM.t(28) 23 Kh IM.m(29) 30 B; IM.s(29) 46 Kh; IM.t(29) 43 Kh IM.t(30) 24 Kh.

DM.#(0) *om.* B; "7 | 21 | 24 | 29" written by a different hand on f. 4ʳ B • DM.#(0) *om.* Kh • Table v *om.* S43 • DM.#(30) *om.* S45 • DM.#(0) *om.* SMB; DM.t(1–30), MM.t(1–12), YM.t(1–20), and IM.t(1–30) *om.* SMB.

Variant IM-table in S45:

30	31	32	33	34	35	36	37	38	39	40	41	42	43	44	45	46	47	48	49	50	51	52	53	54	55	56	57	58	59	60
0	6	0	5	11	5	11	5	10	4	10	4	9	3	9	3	8	2	8	2	7	1	7	1	7	0	6	0	6	11	5
19	12	5	28	21	14	7	0	23	16	8	1	24	17	10	3	26	19	12	5	28	21	14	7	0	23	15	8	1	24	17
34	30	26	23	19	15	12	8	4	1	57	53	50	46	42	38	35	31	27	24	20	16	13	9	5	2	58	54	51	47	43
6	25	45	4	24	43	3	23	42	2	21	41	0	20	39	59	18	38	57	17	36	56	15	35	54	14	33	53	12	32	51
24	55	26	57	27	58	29	0	30	1	32	3	33	4	35	6	36	7	38	9	40	11	41	11	42	13	44	14	45	17	48

Table VI

⌜a f. 6v B
f. 3r Kh
f. 4v S45
f. 3v SMB
⌜b f. 7r B

⌜a‖ buddhaśīghroccadinabhogāḥ ‖ buddhābdabījānivarṣa 22 vikalā 1 nīvṛddhi gatābda 442 | 19 | 444 | 20 | 446 | 21 dhanaṃ ‖

	0	1	2	3	4	⌜b5	6	7	8	9	10	11	12	13	14	15	16	17	18	19	20	21	22	23	24	25	26	27	28	29	30
DM.z	0	0	0	0	0	0	0	0	1	1	1	1	1	1	1	2	2	2	2	2	2	2	3	3	3	3	3	3	3	3	4
DM.d	0	4	8	12	16	20	24	28	2	6	10	15	19	23	27	1	5	9	13	17	21	25	0	4	8	12	16	20	24	28	2
DM.m	0	5	11	16	22	27	33	38	44	49	55	0	6	12	17	23	28	34	39	45	50	56	1	7	12	18	24	29	35	40	46
DM.s	0	32	4	37	9	41	14	46	18	51	23	55	28	0	32	5	37	9	42	14	47	19	51	24	58	48	1	33	5	38	10
DM.t	0	21	42	3	24⌟c	45	6	27	48	9	30	51	12	33	54	15	26	57	18	39	0	21	42	3	28	45	6	27	48	9	30⌟d

⌟c f. 6v B
⌟d f. 3r Kh

⌜e f. 3v Kh
⌜f f. 7v B

‖ buddhaśīghroccamāsāḥ ‖

	⌜e1	2	3	4	5	6	7	8	9	10	11	12
MM.z	4	8	0	4	8	0	4	8	0	4	9	1
MM.d	2	5	8	11	13	16	19	22	24	27	0	3
MM.m	46	32	18	4	50	37	23	9	55	41	27	14
MM.s	10	21	31	42	52	3	13	24	34	45	55	6
MM.t	30	1	32	2	33	3	34	4	35	5	36	6

‖ buddhaśīghroccavarṣabhogāḥ ‖

	1	2	⌜f3	4	5	6	7	8	9	10	11	12	13	14	15	16	17	18	19	20
YM.z	1	2	3	4	5	6	7	8	9	11	0	1	2	3	4	5	6	7	9	10
YM.d	3	6	9	12	16	19	22	25	29	2	5	8	12	15	18	21	24	28	1	4
YM.m	14	28	42	56	10	24	38	52	6	21	35	49	3	17	31	45	59	13	27	42
YM.s	6	12	18	24	30	36	42	48	54	1	7	13	19	25	31	37	43	49	56	2
YM.t	6	13⌟	19	25	32	38	44	51	57	3	10	16	22	29	35	41	48	54	0	7

⌟ f. 7r B

⌜ f. 8r B

‖ buddhaśīghroccakṣepakāḥ ‖

	1	2	3	4	5	6	7	8	9	10	11	12	13	14	15	16	17	18	19	20	21	22	23	24	⌜25	26	27	28	29	30
IM.z	0	11	9	7	5	3	1	11	10	8	6	4	2	0	11	9	7	5	3	1	11	10	8	6	4	2	0	11	9	7
IM.d	25	0	5	10	14	19	24	28	3	8	12	17	22	27	1	6	11	15	20	25	29	4	9	14	18	23	28	2	7	12
IM.m	56	38	20	2	44	26	8	50	32	14	56	38	20	2	45	27	9	51	33	15	57	39	21	3	45	27	9	51	33	15
IM.s	32	34	36	38	40	43	44	48	49	51	53	55	57	59	1	3	5	8	10	12	14	16	18	20	22	24	27	29	31	33
IM.t	7	14	21	20	35	41	48	50	52	39	16	23	30	30	44	17	58	4	11	18	25	32	39	45⌟g	52	59	6	13	20	27⌟h

⌟g f. 7v B
⌟h f. 8r B
f. 3v Kh
f. 4v S45
f. 3v SMB

Table VI: apparatus criticus

‖ buddhaśīghroccadinabhogāḥ°…°121 dhanaṃ ‖] ‖ atha budhabhogaṇī ‖ (f. 6[v]) B; atha budhasya madhyabhogyadināni (f. 3[r]) Kh; buddhoccabhogyadināni ‖ ardhabhukti 122 | 46 ‖ deśāntara 92 vikakalā dhanaṃ ‖ rāmabījaṃ 700 kalā dhanaṃ ‖ SMB • ‖ buddhaśīghroccamāsāḥ ‖] budhamāsāḥ (f. 7[r]) B; budhasya māsāḥ (f. 3[v]) Kh; ‖ buddhoccamāsāḥ ‖ SMB • ‖ buddhaśīghroccavarṣabhogāḥ ‖] budhavarṣāṇiḥ ‖ (f. 7[r]) B; budhamadhyamavarṣāṇi (f. 3[v]) Kh; ‖ buddhoccavarṣāṇi ‖ SMB • ‖ buddhaśīghroccakṣepakāḥ ‖] budhakṣepāḥ 2 | 21 | 14 | 30 (f. 7[r]) B, budhaṣeṣāpūrṇaḥ (f. 8[r]) B; atha budha abdabījāni 22 vilakalā 1 vṛddhiḥ 442 | 19 | 444 | 20 | 446 | 21 dhanaṃ (f. 3[v]) Kh; ‖ budhoccakṣepakāḥ ‖ SMB.

DM.d(1) 59 Kh DM.s(2) 5 SMB DM.s(5) 42 SMB DM.m(7) 36 S45 DM.m(8) 24 B DM.s(8) 19 SMB DM.s(11) 51 Kh, 56 SMB DM.m(13) 11 S45; DM.s(13) 7 Kh DM.m(14) 7 Kh; DM.s(14) 33 SMB; DM.t(14) 5 B DM.m(16) 25 B; DM.t(16) 36 B, Kh DM.s(17) 10 SMB DM.s(19) [-] Kh DM.s(20) 41 B DM.d(22) 3 Kh; DM.m(22) 0 Kh, 21 S45, 52 SMB DM.m(24) 56 B; DM.s(24) 56 Kh, SMB; DM.t(24) 24 B, Kh DM.m(25) [-] S45; DM.s(25) 18 Kh, [-] S45, 29 SMB; DM.t(25) 54 B DM.m(28) 3 SMB; DM.s(28) 6 SMB DM.z(29) 4 Kh.

MM.m(2) 51 B; MM.t(2) 0 Kh MM.t(7) 24 B MM.m(8) 19 B.

YM.t(2) 12 Kh YM.m(4) 16 Kh YM.s(7) 43 SMB YM.s(8) 49 SMB YM.s(9) 55 SMB YM.t(10) 0 B YM.t(11) 6 B YM.t(12) 12 B YM.s(13) 19 SMB; YM.t(13) 18 B YM.t(14) 14 B YM.t(15) 31 B YM.s(16) 38 SMB; YM.t(16) 36 B YM.s(17) 44 SMB; YM.t(17) 42 B YM.s(18) 46 B, 50 SMB; YM.t(18) 48 B YM.t(19) 54 B YM.t(20) 0 B.

IM.s(4) 37 Kh; IM.t(4) 28 Kh IM.s(6) 42 Kh IM.s(7) 45 B IM.m(8) 5 SMB; IM.s(8) 46 Kh, 47 SMB; IM.t(8) 55 Kh IM.t(9) 2 Kh IM.s(10) 52 B; IM.t(10) 9 Kh IM.s(11) 55 B IM.s(12) 57 B IM.s(13) 59 B; IM.t(13) 35 Kh IM.t(14) 37 Kh IM.s(15) 2 SMB; IM.t(15) 45 Kh IM.s(16) 4 SMB; IM.t(16) 51 Kh IM.s(17) 6 SMB IM.t(19) 15 Kh IM.s(24) 21 Kh, SMB; IM.t(24) 42 Kh IM.m(25) 15 SMB; IM.s(25) 23 SMB IM.s(26) 25 SMB IM.s(27) 37 Kh.

DM.#(0) *om.* B; "2 | 21 | 14 | 30" written by a different hand in the right margin of f. 7[v] B • DM.#(0) and DM.s(19) *om.* Kh • Table VI *om.* S43 • DM.#(30) *om.* S45 • DM.#(0) *om.* SMB; DM.t(1–30), MM.t(1–12), YM.t(1–20), and IM.t(1–30) *om.* SMB.

Variant IM-table in S45:

30	31	32	33	34	35	36	37	38	39	40	41	42	43	44	45	46	47	48	49	50	51	52	53	54	55	56	57	58	59	60
7	5	3	1	0	10	8	6	4	2	0	11	9	7	5	3	1	0	10	8	6	4	2	1	11	9	7	5	3	1	0
12	16	21	26	1	5	10	15	19	24	29	3	3	13	18	22	27	2	6	11	16	20	25	0	5	9	14	19	23	28	3
15	57	39	21	3	45	27	9	51	33	15	57	39	22	4	46	28	10	52	34	16	58	40	22	4	46	28	10	52	34	16
33	35	37	39	41	44	46	48	50	52	54	56	58	0	3	5	7	9	11	13	15	17	19	22	24	26	28	30	32	34	36
27	34	41	48	55	2	8	15	22	29	36	43	50	57	4	11	17	24	31	38	45	52	59	6	13	20	26	33	40	47	54

Table VII

⌜ f. 8^{r} B

f. 3^{v} Kh
f. 5^{r} S45
f. 4^{r} SMB

⌜‖ gurudinabhogāḥ ‖ guror abdabījāni gatābdasaṃkhyā iṣṭā ṛṇa gaṃtābdā 442 | 6 | 44 | 7 varṣa 194 antaraṃ tu 1 vṛddhiḥ ‖

	0	1	2	3	4	5	6	7	8	9	10	11	12	13	14	15	16	17	18	19	20	21	22	23	24	25	26	27	28	29	30
DM.z	0	0	0	0	0	0	0	0	0	0	0	0	0	0	0	0	0	0	0	0	0	0	0	0	0	0	0	0	0	0	0
DM.d	0	0	0	0	0	0	0	0	0	0	0	0	0	1	1	1	1	1	1	1	1	1	1	1	1	2	2	2	2	2	2
DM.m	0	4	9	14	19	24	29	34	39	44	49	54	59	4	9	14	19	24	29	34	39	44	49	54	59	4	9	14	19	24	29
DM.s	0	59	58	57	56	55	54	54	53	52	51	50	49	48	48	47	46	45	44	43	42	42	41	40	39	38	37	37	36	35	34
DM.t	0	9	18	27	36	44	53	2	11	20	29	38	47	56	5	14	23	32	41	49	58	7	16	25	34	43	51	0	9	18	27⌟

⌟ f. 3^{v} Kh

⌜a f. 4^{r} Kh
⌜b f. 8^{v} B

‖ gurumāsāḥ ‖

	⌜a1	2	3	4	5	⌜b6	7	8	9	10	11	12
MM.z	0	0	0	0	0	0	0	0	0	0	0	0
MM.d	2	4	7	9	12	14	17	19	22	24	27	29
MM.m	29	59	28	58	27	57	27	56	26	55	25	54
MM.s	34	8	43	17	52	26	1	35	10	44	18	53
MM.t	27	54	21	44	15⌟	42	9	36	3	30	57	24

‖ guruvarṣabhogāḥ ‖

	1	2	3	4	5	6	7	8	9	10	11	12	13	14	15	16	17	18	19	20
YM.z	0	1	2	3	4	5	6	7	8	9	10	11	0	1	2	3	4	5	6	7
YM.d	29	29	29	29	29	29	29	29	29	29	29	28	28	28	28	28	28	28	28	28
YM.m	54	49	44	39	34	29	24	19	14	8	3	58	53	48	43	38	33	28	22	17
YM.s	53	46	40	33	26	20	13	7	0	54	47	40	34	27	21	14	7	1	54	47
YM.t	24	48	12	36	0	24	48	12	36	0	24	48	12	34	0	24	48	12	36	59

⌟ f. 8^{r} B

⌜ f. 9^{r} B

‖ gurukṣepakāḥ ‖

	1	2	3	4	5	6	7	8	9	10	11	12	13	14	15	⌜16	17	18	19	20	21	22	23	24	25	26	27	28	29	30
IM.z	10	6	1	9	5	1	9	5	1	9	5	1	9	5	1	9	5	1	9	4	0	8	4	0	8	4	0	8	4	0
IM.d	2	0	28	27	25	23	22	20	18	16	15	13	11	10	8	6	5	3	1	29	28	26	24	23	21	19	18	16	14	12
IM.m	18	36	54	12	29	47	5	23	41	58	16	34	52	10	27	45	3	21	39	56	14	32	50	8	25	43	1	19	37	54
IM.s	38	26	14	2	50	38	26	14	2	50	38	26	14	2	50	38	26	14	2	50	38	26	14	2	50	38	26	14	2	50
IM.t	59	59	58	58	57	57	56	56	55	55	54	54	53	59	59⌟c	59	59	59	59	50	59	49	58	59	49	49	49	49	49	49⌟d

⌟c f. 8^{v} B
⌟d f. 9^{r} B
f. 4^{r} Kh
f. 5^{r} S45
f. 4^{r} SMB

Table VII: apparatus criticus

‖ gurudinabhogāḥ° ... °1 vṛddhiḥ ‖] ‖ atha gurubhogāḥ ‖ (f. 8r) B; atha gurumadhye bhogyadināni guru abdabījāni gatābdasaṃkhyā iṣṭā ṛṇagatābdāḥ 442 | 6 | 441 7 | varṣa 1947 parantu 1 vṛddhiḥ (f. 3v) Kh; gurubhogyadināni ‖ ardhabhuktiḥ 2 | 30 ‖ rāmabīja 190 kalā ṛṇaṃ ‖ deśāntarakalā 14 dhanaṃ ‖ SMB • ‖ gurumāsāḥ ‖] māsaṣepā (f. 8v) B; atha guror madhyabhogyamāsāḥ ‖ (f. 4r) Kh • ‖ guruvarṣabhogāḥ ‖] guruvarṣa ‖ (f. 8v) B; atha guruvarṣāṇi ‖ (f. 4r) Kh; ‖ guruvarṣāṇi ‖ rāmabīja 190 kalā ṛṇaṃ SMB • ‖ guruṣepakāḥ ‖] gurukṣepāḥ 2 | 4 | 0 | 51 (f. 8v) B, guruṣepāsamāptāḥ (f. 9r) B; atha gurukṣepāḥ ‖ (f. 4r) Kh.

DM.s(5) 56 SMB; DM.t(5) 45 Kh DM.s(6) 55 SMB DM.t(7) 3 B DM.s(12) 50 SMB DM.s(13) 49 SMB DM.d(15) [-] Kh DM.s(16) 45 Kh; DM.t(16) 32 Kh DM.m(17) 29 Kh; DM.s(17) 44 Kh; DM.t(17) 51 Kh DM.m(18) 34 Kh; DM.s(18) 43 Kh, 45 SMB; DM.t(18) 49 Kh DM.m(19) 39 Kh; DM.s(19) 42 Kh, 44 SMB; DM.t(19) 50 B, 58 Kh DM.m(20) 44 Kh; DM.s(20) 42 Kh, 43 SMB; DM.t(20) 7 Kh DM.m(21) 49 Kh; DM.s(21) 41 Kh; DM.t(21) 16 Kh DM.m(22) 54 Kh; DM.s(22) 40 Kh; DM.t(22) 25 Kh DM.m(23) 59 Kh; DM.s(23) 39 Kh; DM.t(23) 34 Kh DM.m(24) 4 Kh; DM.s(24) 38 Kh; DM.t(24) 43 Kh DM.m(25) 9 Kh; DM.s(25) 37 Kh, 39 SMB; DM.t(25) 51 Kh DM.m(26) 14 Kh; DM.s(26) 37 Kh, 38 SMB; DM.t(26) 0 Kh DM.m(27) 19 Kh; DM.s(27) 36 Kh; DM.t(27) 9 Kh DM.t(29) 3̈ Kh.

MM.s(2) 9 SMB MM.s(4) 43 Kh, 18 SMB; MM.t(4) 48 B, 58 Kh; MM.s(6) 27 SMB MM.t(8) 35 Kh MM.m(9) 36 SMB MM.s(11) 19 SMB.

YM.s(2) 47 SMB YM.m(3) 49 B; YM.s(3) 46 B; YM.t(3) 48 B YM.m(5) 24 B; YM.s(5) 17 B, 27 Kh, SMB YM.m(6) 19 SMB; YM.t(6) 20 B, 44 Kh YM.s(7) 14 SMB; YM.t(7) [-]8 Kh YM.t(10) 7 B, 8 Kh YM.m(11) 8 SMB; YM.t(11) 20 B YM.m(12) 56 SMB; YM.s(12) 44 Kh, 41 SMB; YM.t(12) 40 B YM.t(14) 36 B, Kh YM.s(17) 8 SMB YM.s(20) 48 B, SMB; YM.t(20) 0 B.

IM.s(1) 39 SMB IM.s(2) 27 SMB IM.s(3) 15 SMB; IM.t(3) 56 Kh IM.s(4) 9 Kh, 3 SMB IM.s(5) 51 SMB IM.s(6) 39 SMB IM.m(7) 4 B; IM.s(7) 27 SMB IM.m(8) 21 B; IM.s(8) 15 SMB IM.s(9) 3 SMB IM.s(10) 51 SMB IM.s(11) 39 SMB IM.m(12) 38 B; IM.s(12) 27 SMB IM.s(13) 15 SMB IM.s(14) 3 SMB; IM.t(14) 53 Kh IM.s(15) 51 SMB; IM.t(15) 52 Kh IM.d(16) 19 B; IM.s(16) 39 SMB; IM.t(16) 52 Kh IM.s(17) 36 Kh, 27 SMB; IM.t(17) 51 Kh IM.m(18) 31 Kh; IM.s(18) 15 SMB; IM.t(18) 51 Kh IM.m(19) 29 B; IM.s(19) 3 SMB; IM.t(19) 50 Kh IM.s(20) 51 SMB IM.s(21) 39 SMB; IM.t(21) 49 Kh IM.s(22) 27 SMB IM.m(23) 5 SMB; IM.s(23) 15 SMB; IM.t(23) 48 Kh IM.d(24) 29 B; IM.m(24) 0 B; IM.s(24) 3 SMB; IM.t(28) 48 Kh IM.s(25) 20 Kh; IM.t(25) 47 Kh IM.s(26) 39 SMB; IM.t(26) 47 Kh IM.z(27) 8 B; IM.s(27) 27 SMB; IM.t(27) 46 Kh IM.s(28) 15 SMB; IM.t(28) 46 Kh IM.s(29) 3 SMB; IM.t(28) 45 Kh IM.m(30) 58 Kh; IM.s(30) 51 SMB; IM.t(30) 45 Kh.

DM.#(0) *om.* B; MM.m(8): 27 corrected to 56 in the left margin of f. 8v B; "2 | 4 | 0 | 51" written by a different hand on f. 8v B • DM.#(0) and DM.d(15) *om.* Kh; DM.#(16)–DM.#(27) values appear left-shifted for commonly attested values corresponding to arguments 17–28 on f. 3v Kh; DM.#(28), Dm.#(29), Dm.#(30) in the right margin of f. 3v Kh • Table VII *om.* S43 - DM.#(30) *om.* S45 • DM.#(0) *om.* SMB; DM.t(1–30), MM.t(1–12), YM.t(1–20), and IM.t(1–30) *om.* SMB.

Variant IM-table in S45:

30	31	32	33	34	35	36	37	38	39	40	41	42	43	44	45	46	47	48	49	50	51	52	53	54	55	56	57	58	59	60
0	8	4	0	8	4	0	8	3	11	7	3	11	7	3	11	7	3	11	7	3	11	7	3	11	7	2	10	6	2	10
12	11	9	7	6	4	2	0	29	27	25	24	22	20	19	17	15	13	12	10	8	7	5	3	2	0	28	26	25	23	21
54	12	30	48	6	23	41	59	17	35	52	10	28	46	4	21	39	57	15	33	50	8	26	44	2	19	37	55	13	31	48
50	38	26	14	2	50	38	26	14	2	50	38	26	14	2	50	38	26	14	2	50	38	26	14	2	50	38	26	14	2	50
45	44	44	43	43	42	42	41	41	40	40	39	39	38	38	37	37	36	36	35	35	34	34	33	33	32	32	31	31	30	30

Table VIII

⌜ f. 9r B; f. 4r Kh; f. 5v S45; f. 4v SMB

⌜deśāntara 72 vikalā dhanaṃ ‖ śukrabhogyadināni ‖ ardhabhuktiḥ 48 | 4 bhṛgor abdabījāni gatābda 432 | 0 | 46 | 441 | 49 | 45 | 0 | 50 varṣe antaravikalā vṛddhidhanaṃ ‖

	0	1	2	3	4	5	6	7	8	9	10	11	12	13	14	15	16	17	18	19	20	21	22	23	24	25	26	27	28	29	30
DM.z	0	0	0	0	0	0	0	0	0	0	0	0	0	0	0	0	0	0	0	1	1	1	1	1	1	1	1	1	1	1	1
DM.d	0	1	3	4	6	8	9	11	12	14	16	17	19	20	22	24	25	27	28	0	2	3	5	6	8	10	11	13	14	16	18
DM.m	0	36	12	48	24	0	36	12	49	25	1	37	13	49	25	1	38	14	50	26	2	38	14	50	27	3	39	15	51	27	3
DM.s	0	7	15	23	30	38	46	54	1	9	17	25	32	40	48	55	3	11	19	26	34	42	50	57	5	13	21	28	36	44	51
DM.t	0	44	28	12	56	40	24	8	52	36	20	4	48	32	16	57	44	28	12	56	40	24	8	52	36	20	4	48	32	16	55⌟

⌟ f. 9r B

⌜a f. 9v B; ⌜b f. 4v Kh

‖ śukramāsāḥ ‖

	⌜a1	2	3	4	5	6	7	8	9	10	11	12
MM.z	1	3	4	6	8	9	11	0	2	4	5	7
MM.d	18	6	24	12	0	18	6	24	12	0	18	6
MM.m	3	7	11	15	19	23	27	30	34	38	42	46
MM.s	51	43	35	27	19	11	3	55	47	39	31	23
MM.t	55	50	45	40	35	30	24	19	14	9	4	59

‖ rāmabīja 270 kalā× śukravarṣāṇi ‖

	1	2	3	4	5	6	7	8	9	⌜b10	11	12	13	14	15	16	17	18	19	20
YM.z	7	2	9	4	0	7	2	9	5	0	7	2	9	5	0	7	2	10	5	0
YM.d	6	13	20	27	3	10	17	24	0	7	14	21	28	4	11	18	25	1	8	15
YM.m	46	32	19	5	51	38	24	11	57	43	30	16	2	49	35	22	8	54	41	27
YM.s	22	45	8	32	54	17	40	3	26	49	12	35	58	21	49	30	30	53	16	39
YM.t	59	58	57	56	55	54	53	52	50⌟	49	48	47	46	45	44	43	42	41	40	38

⌟ f. 4r Kh

⌜ f. 10r B

‖ śukraśīghroccakṣepakāḥ ‖

	1	2	3	4	5	6	7	8	9	10	11	12	⌜13	14	15	16	17	18	19	20	21	22	23	24	25	26	27	28	29	30
IM.z	9	9	10	10	11	11	0	0	1	1	2	2	3	3	4	4	5	5	6	6	7	7	8	8	9	10	10	11	11	0
IM.d	3	19	4	19	5	20	6	21	7	22	8	23	9	24	10	25	10	26	11	27	12	28	13	29	14	0	15	1	16	1
IM.m	33	1	28	56	24	51	19	47	14	42	10	37	5	33	0	28	56	33	51	19	46	14	42	9	37	5	32	0	28	55
IM.s	34	14	53	33	13	52	32	12	51	31	11	50	30	10	49	29	8	48	58	7	47	27	6	46	26	5	45	25	4	44
IM.t	39	17	17	33	13	51	30	8	47	26	4	43⌟a	21	47	39	17	56	13	13	51	30	9	47	36	4	43	21	0	38	17⌟b

⌟a f. 9v B; ⌟b f. 10r B; f. 4v Kh; f. 5v S45; f. 4v SMB

Table VIII: apparatus criticus

‖ deśāntara 72 vikalādhanaṃ°…°vṛddhidhanaṃ ‖] ‖ atha śukrabhogāḥ ‖ (f. 9r) B; atha bhṛgumadhyabhogyadināni abdabījāni gatābda 432 | 0 | 46 | 441 | 49 | 450 | 5 varṣantarantarāḥ kalāvṛddhiḥ[-] dhanam [-] (f. 4r) Kh; śukraśīghroccabhogyadināni ‖ SMB • ‖ śukramāsāḥ ‖] śukramāśāḥ (f. 9v) B; bhṛgumadhyamāsāḥ (f. 4r) Kh; ‖ śukraśīghroccamāsāḥ ‖ SMB • ‖ rāmabīja 270 kalā× śukravarṣāṇi ‖] śukravarṣāṇiḥ ‖ (f. 9v) B; bhṛguvarṣāṇi ‖ (f. 4r) Kh; ‖ śukraśīghroccavarṣāṇi ‖ rāmabījaṃ 270 × SMB • ‖ śukraśīghroccakṣepakāḥ ‖] śukramadhyakṣepā 8 | 18 | 5 | 55 (f. 9v) B; bhṛgukṣepāḥ (f. 4v) Kh.

DM.z(1) 1 Kh; DM.s(1) 8 SMB DM.t(2) 27 Kh DM.m(3) 28 B; DM.s(3) 22 B; DM.t(3) 11 Kh DM.m(4) 40 Kh; DM.s(4) 31 SMB; DM.t(4) 55 B, 52 Kh DM.t(5) 39 B, Kh DM.s(6) 46 SMB; DM.t(6) 23 B, Kh DM.t(7) 7 B, Kh DM.s(8) 2 SMB; DM.t(8) 51 B, 50 Kh DM.t(9) 34 B, 24 Kh DM.t(10) 19 B, 18 Kh DM.s(11) 20 B; DM.t(11) 2 B, Kh DM.d(12) 18 B; DM.s(12) 33 SMB; DM.t(12) 46 B, Kh DM.t(13) 31 B, 9 Kh DM.t(14) 15 B, 12 Kh DM.s(15) 56 SMB; DM.t(15) 58 B DM.s(16) 4 SMB; DM.t(16) 43 B, 47 Kh DM.t(17) 27 B, 25 Kh DM.t(18) 9 B, Kh DM.s(19) 27 SMB; DM.t(19) 49 B, 52 Kh DM.t(20) 39 B, 36 Kh DM.t(21) 23 B, 20 Kh DM.t(22) 7 B, 4 Kh DM.s(23) 58 SMB; DM.t(23) 51 B, 48 Kh DM.t(24) 35 B, 32 Kh DM.t(25) 19 B, 18 Kh DM.s(26) 36 SMB; DM.t(26) 3 B, 0 Kh DM.m(27) 25 S45; DM.s(27) 29 SMB, 18 Kh; DM.t(27) 47 B, 44 Kh DM.t(28) 31 B, 27 Kh DM.t(29) 15 B, 11 Kh DM.s(30) 52 SMB; DM.t(30) 59 B.

MM.m(1) 36 SMB; MM.s(1) 52 SMB MM.s(2) 44 SMB MM.m(3) 44 SMB; MM.s(3) 36 SMB MM.z(4) 0 SMB; MM.d(4) 6 SMB; MM.m(4) 24 SMB; MM.s(4) 31 SMB MM.s(5) 19 SMB MM.d(10) 7 Kh MM.s(11) 42 S45 MM.s(12) 22 Kh, S45; MM.t(12) 0 B.

YM.s(1) 23 SMB; YM.t(1) 58 B YM.m(2) 42 SMB; YM.s(2) 36 SMB YM.s(3) 9 B YM.s(4) 31 B, Kh YM.s(5) 55 SMB YM.s(6) 18 SMB YM.m(7) 34 Kh; YM.s(7) 41 SMB YM.s(8) 4 SMB; YM.t(8) 51 B, Kh YM.s(9) 27 SMB YM.s(10) 50 SMB YM.s(11) 13 SMB YM.s(12) 36 SMB YM.s(13) 59 SMB YM.s(14) 22 SMB YM.s(15) 44 B, Kh, 45 SMB YM.s(16) 7 B, Kh, 8 SMB YM.s(17) 31 SMB YM.s(18) 54 SMB; YM.t(18) 40 B YM.s(19) 17 SMB; YM.t(19) 38 B YM.s(20) 29 B, 31 Kh; YM.t(20) 34 B.

IM.s(3) 54 SMB; IM.t(3) 56 Kh IM.t(4) 34 Kh IM.s(6) 53 SMB IM.s(9) 52 SMB IM.z(11) 1 Kh IM.m(12) 27 SMB; IM.s(12) 51 SMB IM.t(13) 11 Kh IM.t(14) 0 Kh IM.d(16) 15 SMB IM.s(17) 9 SMB IM.m(18) 23 B; IM.t(18) 34 B, Kh IM.s(19) 28 SMB IM.s(20) 8 SMB IM.s(23) 7 SMB; IM.t(23) 40 Kh IM.t(24) 26 Kh IM.d(25) 24 Kh IM.d(26) 14 Kh; IM.m(26) 35 Kh; IM.s(26) 45 Kh; IM.t(26) 1 Kh; IM.t(29) 34 Kh.

DM.#(0) *om.* B; "8 | 18 | 5 | 55" written by a different hand on f. 9v B • DM.#(0) *om.* Kh; YM-table has an additional entry [0 | 8 | 18 | 5 | 55 | 9] on f. 4v Kh • Table VIII *om.* S43 • DM.#(30) *om.* S45 • DM.#(0) *om.* SMB; DM.t(1–30), MM.t(1–12), YM.t(1–20), and IM.t(1–30) *om.* SMB.

Variant IM-table in S45:

30	31	32	33	34	35	36	37	38	39	40	41	42	43	44	45	46	47	48	49	50	51	52	53	54	55	56	57	58	59	60
0	0	1	1	2	2	3	3	4	4	5	5	6	6	7	7	8	8	9	9	10	10	11	11	0	0	1	1	2	3	3
1	17	2	18	3	19	4	20	5	21	6	22	7	22	8	23	9	24	10	25	11	26	12	27	12	28	13	29	14	0	15
55	23	51	18	46	14	41	9	37	4	32	0	27	55	22	50	18	45	13	41	8	36	4	31	59	27	54	22	50	17	45
44	23	3	43	22	2	42	21	1	41	20	0	39	19	59	38	18	58	37	17	57	36	16	46	35	15	54	34	14	53	33
17	55	34	12	51	29	8	46	25	3	42	20	59	37	16	54	33	11	50	28	7	45	24	2	41	19	58	36	15	53	32

Table IX

⌜a f. 10r B; f. 4v Kh; f. 6r S45; f. 5r SMB; ⌜b f. 10v B; ⌟ f. 10r B

	⌜adeśāntara o dhanaṃ ‖ śanibhogyadināni ‖ ardhabhuktiḥ 1 | o ‖ śrīḥ ‖																														
	0	1	2	3	4	5	6	7	8	9	10	11	12	13	14	15	16	17	18	19	20	21	22	23	24	25	26	27	⌜b28	29	30
DM.z	0	0	0	0	0	0	0	0	0	0	0	0	0	0	0	0	0	0	0	0	0	0	0	0	0	0	0	0	0	0	0
DM.d	0	0	0	0	0	0	0	0	0	0	0	0	0	0	0	0	0	0	0	0	0	0	0	0	0	0	0	0	0	0	1
DM.m	0	2	4	6	8	10	12	14	16	18	20	22	24	26	28	30	32	34	36	38	40	42	44	46	48	50	52	54	56	58	0
DM.s	0	0	0	1	1	1	2	2	3	3	3	4	4	5	5	5	6	6	6	7	7	8	8	8	9	9	10	10	10	11	11
DM.t	0	23	46	9	32	55	18	41	4	17	50	13	36	0	23	46	9	32	55	18	41	4	27	27	13	36	0	23⌟	46	9	33

	‖ śanimāsāḥ ‖											
	1	2	3	4	5	6	7	8	9	10	11	12
MM.z	0	0	0	0	0	0	0	0	0	0	0	0
MM.d	1	2	3	4	5	6	7	8	9	10	11	12
MM.m	0	0	0	0	0	1	1	1	1	1	2	2
MM.s	11	23	34	46	57	9	20	32	43	55	6	18
MM.t	33	3	35	7	39	11	43	14	46	10	2	21

	‖ rāmabīja 30 dhanaṃ śanivarṣāṇi ‖																			
	1	2	3	4	5	6	7	8	9	10	11	12	13	14	15	16	17	18	19	20
YM.z	0	0	1	1	2	2	2	3	3	4	4	4	5	5	6	6	6	7	7	8
YM.d	12	24	6	18	0	12	24	6	18	0	12	24	6	18	0	12	24	6	18	0
YM.m	2	4	6	9	11	13	16	18	20	23	25	27	29	32	34	36	39	41	43	46
YM.s	18	36	55	13	31	50	8	26	45	3	21	40	58	17	35	53	12	30	48	7
YM.t	21	42	4	26	46	9	30	52	13	35	56	18	39	0	22	43	50	26	48	9

⌜a f. 5r Kh; ⌜b f. 11r B; ⌟c f. 4v Kh; ⌟d f. 10v B; ⌟e f. 11r B; f. 5r Kh; f. 6r S45; f. 5r SMB

	‖ śanikṣepakāḥ ‖																													
	1	2	3	4	5	6	7	8	⌜a9	10	11	12	13	⌜b14	15	16	17	18	19	20	21	22	23	24	25	26	27	28	29	30
IM.z	0	8	4	0	8	4	0	8	4	0	8	4	0	8	4	0	8	4	0	8	4	0	8	4	0	8	4	0	8	4
IM.d	4	5	6	6	7	8	9	9	10	11	12	12	13	14	15	16	16	17	18	19	19	20	21	22	22	23	24	25	26	26
IM.m	29	15	1	47	33	19	6	52	38	24	10	56	42	28	15	1	47	33	19	5	51	37	24	10	56	42	28	14	0	46
IM.s	24	31	38	45	52	59	7	14	21	28	35	42	49	57	4	11	19	25	33	40	47	54	1	8	16	23	30	37	44	51
IM.t	10	19	29	39	48	58	8	17⌟c	27	37	46	56	16⌟d	15	25	35	35	54	52	14	23	33	53	52	2	15	21	31	41	50⌟e

Table IX: apparatus criticus

‖ deśāntara o dhanaṃ°…°so śrīḥ ‖] ‖ śanibhaugāḥ ‖ (f. 10^{r}) B; atha śaner madhyabhogyadināni (f. 4^{v}) Kh; śanibhogyadināni ‖ ardhabhukti ‖ deśāntara dhanaṃ ‖ SMB • ‖ śanimāsāḥ ‖] śaneḥ masāḥ (f. 4^{v}) Kh; ‖ • ‖ rāmabīja 30 dhanaṃ śanivarṣāṇi ‖] śanivarṣāṇiḥ ‖ (f. 10^{v}) B; śaner varṣāṇi (f. 4^{v}) Kh; ‖ śanivarṣāṇi ‖ SMB • ‖ śanikṣepakāḥ ‖] śaniṣepāḥ | 4 | 3 | 43 | 17 (f. 10^{v}) B; śanekṣepāḥ (f. 4^{v}), śaneḥ kṣepāḥ ‖ (f. 5^{r}) Kh.

DM.s(2) 1 SMB DM.m(4) 10 SMB; DM.s(4) 0 SMB DM.s(5) 2 SMB; DM.t(5) 52 Kh DM.s(7) 3 SMB; DM.t(7) 47 B DM.t(9) 27 B, Kh DM.s(10) 4 SMB; DM.t(10) 27 B DM.s(13) 4 Kh; DM.t(13) 59 Kh DM.s(15) 6 SMB DM.t(16) 9 B DM.s(19) 6 B; DM.t(19) 55 B DM.s(20) 8 SMB; DM.t(20) 18 B DM.s(21) 7 B; DM.t(21) 41 B DM.t(22) 4 B DM.s(23) 9 SMB; DM.t(23) 47 B, 50 Kh DM.s(24) 8 B; DM.t(24) 50 B DM.t(26) 50 B DM.s(27) 11 B DM.s(28) 11 B, SMB; DM.t(28) 44 B DM.s(29) 12 B; DM.t(29) 7 B DM.d(30) 0 SMB; DM.s(30) 12 B; DM.t(30) 32 B, Kh.

MM.t(1) 32 Kh MM.t(2) 4 Kh MM.t(3) 36 Kh MM.t(4) 8 Kh MM.t(5) 40 Kh MM.d(6) 5 Kh; MM.t(6) 12 Kh MM.d(7) 6 Kh; MM.s(7) 21 SMB; MM.t(7) 35 Kh MM.d(8) 7 Kh; MM.t(8) 7 Kh MM.d(9) 8 Kh; MM.s(9) 53, Kh, 44 SMB MM.t(9) 39 Kh MM.d(10) 9 Kh; MM.m(10) 2 Kh; MM.t(10) 5 Kh; MM.t(10) 18 B, 11 Kh MM.d(11) 10 Kh MM.s(11) 16 Kh, 7 SMB; MM.t(11) 50 B, 43 Kh MM.d(12) 11 Kh; MM.s(12) 28 Kh; MM.t(12) 27 B, 15 Kh.

YM.s(2) 37 SMB; YM.t(2) 43 B YM.s(5) 32 SMB; YM.t(5) 47 B, Kh YM.s(7) 6 SMB YM.s(8) 27 SMB YM.s(9) 15 S45 YM.s(11) 22 SMB YM.m(13) 19 B YM.s(14) 16 B; YM.t(14) 54 B YM.t(15) 25 B, 27 Kh YM.s(16) 57 S45, 59 SMB; YM.t(16) 36 B YM.t(17) 57 B, 5 Kh YM.m(18) 42 SMB; YM.t(18) 18 B YM.s(19) 49 Kh, SMB; YM.t(19) 38 B YM.z(20) 0 Kh; YM.d(20) 4 Kh; YM.m(20) 29 Kh; YM.s(20) 24 Kh; YM.t(20) 10 Kh.

IM.z(1) 8 Kh; IM.d(1) 5 Kh; IM.m(1) 15 Kh; IM.s(1) 31 Kh; IM.t(1) 19 Kh IM.z(2) 4 Kh; IM.d(2) 6 Kh; IM.m(2) 1 Kh; IM.s(2) 38 Kh; IM.t(2) 29 Kh IM.z(3) 0 Kh; IM.m(3) 47 Kh; IM.s(3) 4 Kh; IM.t(3) 0 Kh IM.d(4) 7 B IM.s(5) 53 SMB IM.m(6) 29 B; IM.t(6) 56 Kh IM.m(7) 19 SMB IM.s(8) 19 SMB IM.t(9) 17 Kh IM.s(11) 36 SMB IM.s(12) 43 SMB IM.m(13) 52 Kh; IM.s(13) 50 Kh, SMB; IM.t(13) 6 Kh IM.m(16) 11 Kh; IM.t(16) 34 Kh IM.m(17) 1 B; IM.s(17) 11 B, 18 Kh; IM.t(17) 35 B, 44Kh IM.s(18) 26 SMB IM.s(19) 4 Kh IM.d(22) 30 Kh IM.m(23) 14 Kh; IM.s(23) 2 SMB; IM.t(23) 43 B, 23 Kh IM.s(24) 9 SMB IM.t(26) 12 Kh IM.m(29) 10 Kh; IM.s(29) 45 SMB IM.s(30) 52 SMB.

DM.#(0) *om.* B; "4 | 3 | 43 | 17" written by a different hand on f. 10^{v} B • DM.#(0) *om.* Kh; the table MM.#(1–12) is transposed and placed after the YM-table on f. 4^{v} Kh • Table IX *om.* S43 • DM.#(30) *om.* S45 • DM.#(0) *om.* SMB; DM.t(1–30), MM.t(1–12), YM.t(1–20), and IM.t(1–30) *om.* SMB.

Variant IM-table in S45:

30	31	32	33	34	35	36	37	38	39	40	41	42	43	44	45	46	47	48	49	50	51	52	53	54	55	56	57	58	59	60
4	0	8	4	0	9	5	1	9	5	1	9	5	1	9	5	1	9	5	1	9	5	1	9	5	1	9	5	1	9	5
26	27	28	29	29	0	1	2	2	3	4	5	6	6	7	8	9	9	10	11	12	12	13	14	15	15	16	17	18	19	19
46	32	19	5	51	37	23	9	55	41	28	14	0	46	32	18	4	50	37	23	9	55	41	27	13	59	45	32	18	4	50
51	59	6	13	20	27	34	41	49	56	3	10	17	24	32	39	46	53	0	7	15	22	29	36	43	50	58	5	12	19	26
51	1	11	20	30	40	49	59	9	18	28	38	48	57	7	17	26	36	46	55	5	15	25	34	44	54	3	13	23	32	42

Table x

⌜‖ mandaphalaṃ adho'ntaraṃ tad adho gatiphalaṃ ‖ ravimandaphalāni ‖ adho gatiphalaṃ ‖ ravimandoccaṃ 2 | 18 | 0 | 0 kendravaśādhanarṇaṃ ‖

⌜ f. 11^{r} B; f. 5^{r} Kh; f. 1^{v} S43; f. 6^{v} S45

	1	2	3	4	5	6	7	8	9	10	11	12	13	14	15	16	17	18	19	20	21	22	23	24	25	26	27	28	29	30
EC.d	0	0	0	0	0	0	0	0	0	0	0	0	0	0	0	0	0	0	0	0	0	0	0	0	0	0	0	1	1	1
EC.m	2	4	6	9	11	13	16	18	20	22	25	27	29	31	33	36	38	40	42	44	46	48	50	53	55	57	59	1	3	5
EC.s	17	34	52	9	27	44	2	19	37	54	5	16	27	38	49	0	10	21	32	43	48	52	56	1	5	9	14	18	22	27
DF.m	2	2	2	2	2	2	2	2	2	2	2	2	2	2	2	2	2	2	2	2	2	2	2	2	2	2	2	2	2	1
DF.s	17	18	17	18	17	18	17	18	17	11	11	11	11	11	11	11	11	11	11	4	4	4	5	4	4	5	4	4	5	51
ID.m	2	2	2	2	2	2	2	2	2	2	2	2	2	2	2	2	2	2	2	2	2	2	2	2	2	2	2	2	2	1
ID.s	20	20	20	20	20	20	20	20	20	13	13	13	13	13	13	13	13	13	13	6	6	6	6	6	6	6	6	6	6	53⌟

⌟ f. 11^{r} B

⌜ f. 11^{v} B

	⌜31	32	33	34	35	36	37	38	39	40	41	42	43	44	45	46	47	48	49	50	51	52	53	54	55	56	57	58	59	60
EC.d	1	1	1	1	1	1	1	1	1	1	1	1	1	1	1	1	1	1	1	1	1	1	1	1	1	1	1	1	1	1
EC.m	7	9	11	12	14	16	18	20	22	24	25	27	28	30	32	33	35	37	38	40	41	42	44	45	46	48	49	50	52	53
EC.s	18	9	1	52	43	34	26	17	8	0	38	16	54	32	10	49	27	5	43	21	40	58	17	36	54	13	31	50	8	27
DF.m	1	1	1	1	1	1	1	1	1	1	1	1	1	1	1	1	1	1	1	1	1	1	1	1	1	1	1	1	1	0
DF.s	51	52	51	51	51	52	51	51	52	38	38	38	38	38	39	38	38	38	38	19	18	19	19	19	19	18	19	18	19	59
ID.m	1	1	1	1	1	1	1	1	1	1	1	1	1	1	1	1	1	1	1	1	1	1	1	1	1	1	1	1	1	1
ID.s	53	53	53	53	53	53	53	53	53	40	40	40	40	40	40	40	40	40	40	20	20	20	20	20	20	20	20	20	20	0⌟

⌟ f. 1^{v} S43

⌜a f. 2^{r} S43; ⌜b f. 5^{v} Kh; ⌜c f. 12^{r} B

	⌜a61	62	63	64	65	66	67	68	⌜b69	70	71	72	73	74	75	76	77	78	⌜c79	80	81	82	83	84	85	86	87	88	89	90
EC.d	1	1	1	1	1	1	2	2	2	2	2	2	2	2	2	2	2	2	2	2	2	2	2	2	2	2	2	2	2	2
EC.m	54	55	56	57	58	59	0	1	2	3	3	4	4	5	6	6	7	7	8	8	8	9	9	9	9	10	10	10	10	10
EC.s	26	25	24	22	21	20	19	18	17	16	49	21	54	27	0	32	5	38	10	43	56	9	22	35	49	2	15	28	41	54
DF.m	0	0	0	0	0	0	0	0	0	0	0	0	0	0	0	0	0	0	0	0	0	0	0	0	0	0	0	0	0	0
DF.s	59	59	58	59	59	59	59	59	59	33	32	33	33	33	33	33	33	32	33	13	13	13	13	13	13	13	13	13	13	0
ID.m	1	1	1	1	1	1	1	1	1	0	0	0	0	0	0	0	0	0	0	0	0	0	0	0	0	0	0	0	0	0
ID.s	0	0	0	0	0	0	0	0⌟d	0	33	33	33	33	33	33	33	33	33⌟e	33	13	13	13	13	13	13	13	13	13	13	13⌟f

⌟d f. 5^{r} Kh; ⌟e f. 11^{r} B; ⌟f f. 12^{r} B; f. 5^{v} Kh; f. 2^{r} S43; f. 6^{v} S45

Table x: apparatus criticus

‖ mandaphalaṃ°] ravyādīnāmandaphalaṃ ‖ mandoccasūrya 2 | 18 | 0 | 0 | (f. 11r) B; atha ravimandaphalam ‖ (f. 5r) Kh, ravimandaphalaṃ saṃpūrṇam (f. 5v) Kh; ‖ atha raver aṃśādinimandaphalāni ‖ te mandaphalajoyānoprakārasvadeśīsūryane sūryanomandoccamānthākā[-]ā itemṛdukendrathāitehanābhujāṃśa uparemandaphalasāntaraṃgrahāi antarekarābhujanukalādikanegomūtrākāigaṇāphalalādhataṃ koṣṭakaṃbhāṃ eṣyapaṅkti adhikahoyatodhanakājeṃ[-]nahoyato ṛṇakājeta mandaphalaspaṣṭathāi imaravyādikasarvegrahanā mandaphalakāje (f. 1v) S43, ‖ raver mandaphalāni ‖ svarṇaṃ phalaṃ meṣatulādikendraṃ ityanena sarvatra dharṇaṃ jñeyaṃ ‖ (f. 2r) S43.

EC.s(3) 51 B   EC.m(7) 19 Kh   DF.s(8) 11 S43   DF.s(9) 11 S43   EC.s(12) 26 S45   EC.s(14) 34 S45   EC.m(16) 16 Kh; EC.s(16) 34 S45; DF.s(16) 10 S45   EC.s(17) 11 S43   EC.m(18) 41 Kh; EC.s(18) 22 S43   EC.s(19) 33 S43   EC.s(20) 41 B, 44 S43; DF.s(20) 44 S45   DF.s(23) 4 S43   DF.s(24) 5 S43   EC.s(25) 3 B   DF.s(26) 4 S43   DF.s(27) 5 S43   EC.m(29) 32 Kh; DF.s(29) 4 S43   EC.m(30) 7 B; DF.s(30) 5 S43; ID.m(30) 2 B   DF.s(36) 51 S43   DF.s(37) 52 S45   EC.s(38) 27 B   DF.s(45) 38 S43   EC.s(48) 27 S43   DF.s(49) 39 S43   EC.s(50) 27 B   DF.s(51) 19 S43   EC.s(52) 57 B   EC.s(53) 16 B, 7 Kh   DF.s(54) 18 S43, S45   EC.s(55) 53 B   EC.s(56) 53 B; DF.s(56) 19 S43   EC.s(57) 32 S43   EC.s(58) 51 B, 49 Kh, 51 S43; DF.s(58) 19 S43   EC.s(59) 10 S43   EC.s(60) 56 B; ID.s(60) 20 Kh   EC.s(61) 27 Kh   DF.s(63) 59 S43, S45   EC.s(64) 23 S43   EC.s(65) 22 S43   EC.s(66) 2 Kh, 51 S43   EC.s(67) 29 S43   DF.s(68) 33 S43   DF.s(69) 33 S43; ID.m(69) 0 Kh; ID.s(69) 33 Kh   DF.s(71) 33 S43, S45   EC.s(72) 34 Kh   DF.s(78) 33 S43, S45   ID.s(79) 32 Kh   DF.s(80) 23 S43; ID.s(80) 33 B   EC.s(85) 59 B.

DF.#(1–90) *om.* B • DF.#(1–90) *om.* Kh • "bhujarāśi 30 guṇokarī aṃśayuktakarīneḥ āka uparīṃ phalajomvūṃ phalaheve antarabai antarathībhujaṃ nākalāvikalā āthemgaṇīkalādikaṃ phaladhanaḥ mandaphalamadhye devūṃtyāremandaphalasphaṣṭaṃ ‖" vernacular paratext in the left margin on f. 6v S45 • Table x *om.* SMB.

Table XI

⌜ f. 11^{r} B, f. 5^{v} Kh, f. 2^{r} S43, f. 7^{r} S45

⌜‖ candramandaphalāni ‖

	1	2	3	4	5	6	7	8	9	10	11	12	13	14	15	16	17	18	19	20	21	22	23	24	25	26	27	28	29	30
EC.d	0	0	0	0	0	0	0	0	0	0	0	1	1	1	1	1	1	1	1	1	1	1	1	2	2	2	2	2	2	2
EC.m	5	10	15	21	26	31	37	42	47	52	57	3	8	13	18	23	28	33	38	43	48	52	57	2	7	12	16	21	26	31
EC.s	17	35	52	10	28	45	3	21	38	56	59	1	4	6	9	11	14	16	19	21	9	56	44	31	18	6	53	40	28	15
DF.m	5	5	5	5	5	5	5	5	5	5	5	5	5	5	5	5	5	5	5	4	4	4	4	4	4	4	4	4	4	4
DF.s	18	17	18	18	18	18	18	18	18	3	2	3	3	3	2	3	3	3	2	48	47	47	47	47	48	47	47	47	48	17
ID.m	68	68	68	68	68	68	68	68	68	65	65	65	65	65	65	65	65	65	65	61	61	61	61	61	61	61	61	61	61	55
ID.s	15	15	15	15	15	15	15	15	15	0	0	0	0	0	0	0	0	0	0	45	45	45	45	45	45	45	45	45	45	15

⌜ f. 12^{v} B

	31	32	33	34	35	⌜36	37	38	39	40	41	42	43	44	45	46	47	48	49	50	51	52	53	54	55	56	57	58	59	60
EC.d	2	2	2	2	2	2	3	3	3	3	3	3	3	3	3	3	3	3	3	3	3	3	4	4	4	4	4	4	4	4
EC.m	35	39	44	48	52	56	1	5	9	14	17	21	25	29	33	36	40	44	48	51	54	57	1	4	7	10	13	16	19	22
EC.s	32	50	7	24	41	58	15	32	49	7	53	40	27	14	1	48	35	22	9	56	57	59	0	2	3	5	6	8	9	11
DF.m	4	4	4	4	4	4	4	4	4	3	3	3	3	3	3	3	3	3	3	3	3	3	3	3	3	3	3	3	3	2
DF.s	16	17	17	17	17	18	17	17	18	46	47	47	47	47	47	47	47	47	47	1	2	1	2	1	2	1	2	1	2	16
ID.m	55	55	55	55	55	55	55	55	55	48	48	48	48	48	48	48	48	48	48	39	39	39	39	39	39	39	39	39	39	29
ID.s	15	15	15	15	15⌟a	15	15	15	15	45	45	45	45	45	45	45	45	45	45	0	0	0	0	0	0	0	0	0	0	15⌟b

⌟a f. 12^{r} B
⌟b f. 2^{r} S43

⌜c f. 2^{v} S43
⌜d f. 6^{r} Kh
⌜e f. 13^{r} B

	⌜c61	62	63	64	65	66	67	68	⌜d69	70	71	72	73	74	75	76	77	78	79	80	81	82	83	⌜e84	85	86	87	88	89	90
EC.d	4	4	4	4	4	4	4	4	4	4	4	4	4	4	4	4	4	4	4	4	4	4	4	4	4	5	5	5	5	5
EC.m	24	26	28	31	33	35	38	40	42	44	46	47	48	49	51	52	53	54	56	57	57	58	58	59	59	0	1	1	2	2
EC.s	27	43	59	15	31	47	4	20	36	52	8	23	39	54	10	26	41	57	13	28	58	29	59	29	59	31	0	30	1	31
DF.m	2	2	2	2	2	2	2	2	2	2	1	1	1	1	1	1	1	1	1	0	0	0	0	0	0	0	0	0	0	0
DF.s	16	16	16	16	16	16	16	16	16	16	15	16	15	16	16	15	16	16	15	30	31	30	30	30	31	30	30	30	30	0
ID.m	29	29	29	29	29	29	29	29	29	16	16	16	16	16	16	16	16	16	16	6	6	6	6	6	6	6	6	6	6	6
ID.s	15	15	15	15	15	15	15	15⌟f	15	15	15	15	15	15	15	15	15	15	15	30	30	30	30	30	30	30	30	30	30	30⌟g

⌟f f. 5^{v} Kh
⌟g f. 13^{r} B, f. 6^{r} Kh, f. 2^{v} S43, f. 7^{r} S45

Table XI: apparatus criticus

|| candramandaphalāni ||] || atha candramandaḥ (f. 12r) B, candramandasamāpatāḥ || (f. 13r) B; candramandaphalādhaḥ gatiphalam anyadeśeṣaṭkarma ete paścāt uccamadhyetā[-]tiṇekendrakalyate meṣatvāt pl alāditvāt pacheṃteha uccapa[-]aṃpūrvetehana ubhujakarīkoṣṭakalevauḥ (f. 5v) Kh; atha candrasya mandaphalāny āṃśādīnāḥ || (f. 2r) S43, candrasya mandaphalāni || (f. 2v) S43.

EC.d(5) 2 S45; DF.s(5) 17 S43 DF.s(8) 17 S43 EC.s(9) 22 B DF.s(13) 2 S43 EC.s(15) 19 Kh EC.s(16) 23 Kh DF.s(17) 2 S43 EC.m(18) 13 Kh EC.m(20) 47 S45 DF.s(22) 48 S43 EC.s(23) 43 B, Kh, S43; DF.s(23) 48 S43 EC.s(24) 21 B; DF.s(24) 47 S43 DF.s(28) 48 S43 DF.s(29) 47 S43 DF.s(31) 18 S43 EC.s(35) 48 B DF.s(36) 17 S43 EC.m(37) 11 Kh EC.s(38) 34 B EC.s(41) 3 S45 EC.s(42) 3 S45 EC.s(43) 3 S45 EC.s(44) 10 Kh, 3 S45 EC.s(45) 3 S45 EC.s(46) 49 Kh, 3 S45 EC.s(47) 55 B, 3 S45 EC.m(48) 45 B; EC.s(48) 23 S45; EC.s(48) 3 S45 EC.s(49) 3 S45 EC.s(50) 55 B EC.m(51) 57 B; EC.s(51) 59 B DF.s(52) 2 S43 EC.s(55) 6 Kh EC.s(56) 1 S45. EC.s(59) 8 Kh ID.m(60) 39 Kh; ID.s(60) 0 Kh EC.s(64) 25 S45 DF.s(66) 17 S43 EC.s(67) 3 S43 ID.m(68) 69 S43 ID.m(69) 39 B, 69 S43 EC.s(70) 51 Kh; ID.m(70) 39 B ID.m(71) 39 B EC.s(72) 26 Kh; ID.m(72) 39 B EC.m(79) 58 Kh EC.s(80) 58 B EC.m(81) 58 B EC.s(82) 25 B EC.m(83) 58 B EC.s(86) 3 B, 30 Kh, S43 DF.s(88) 31 S43 EC.s(89) 0 Kh EC.s(90) 30 Kh; ID.m(90) 68 B; ID.s(90) 25 B.

DF.#(1–90) *om.* B • DF.#(1–90) *om.* Kh • EC.s(38) 35^2 correction on f. 2r S43 • Table XI *om.* SMB.

Table XII

⌜ f. 13ᵛ B, f. 6ʳ Kh, f. 2ᵛ S43, f. 8ʳ S45

⌜‖ bhaumamandaphalāni ‖

	1	2	3	4	5	6	7	8	9	10	11	12	13	14	15	16	17	18	19	20	21	22	23	24	25	26	27	28	29	30
EC.d	0	0	0	0	0	1	1	1	1	1	2	2	2	2	2	3	3	3	3	3	4	4	4	4	4	4	5	5	5	5
EC.m	11	23	35	47	58	10	22	34	45	57	8	20	31	42	53	5	16	27	38	49	0	11	21	32	43	53	4	15	25	36
EC.s	46	33	19	6	52	39	25	12	58	45	58	11	24	37	49	2	15	28	41	54	33	12	52	31	10	49	29	8	47	26
DF.m	11	11	11	11	11	11	11	11	11	11	11	11	11	11	11	11	11	11	11	10	11	10	10	10	10	10	10	10	10	9
DF.s	47	46	47	47	47	46	47	46	47	13	13	13	13	12	13	13	13	13	13	39	39	40	39	39	39	40	39	39	39	32
ID.m	6	6	6	6	6	6	6	6	6	5	5	5	5	5	5	5	5	5	5	5	5	5	5	5	5	5	5	5	5	4
ID.s	0	0	0	0	0	0	0	0	0	43	43	43	43	43	43	43	43	43	43	25	25	25	25	25	25	25	25	25	25	51⌟

⌟ f. 2ᵛ S43

⌜a f. 3ʳ S43, ⌜b f. 14ʳ B

	⌜a31	32	33	34	35	36	⌜b37	38	39	40	41	42	43	44	45	46	47	48	49	50	51	52	53	54	55	56	57	58	59	60
EC.d	5	5	6	6	6	6	6	6	7	7	7	7	7	7	7	8	8	8	8	8	8	8	8	9	9	9	9	9	9	9
EC.m	45	55	5	14	24	33	43	52	2	11	20	28	37	45	53	2	10	19	27	35	42	49	56	2	9	16	22	29	36	43
EC.s	58	30	2	34	6	38	10	42	14	46	11	35	0	25	49	14	39	3	28	53	37	20	4	48	31	15	59	43	26	10
DF.m	9	9	9	9	9	9	9	9	9	8	8	8	8	8	8	8	8	8	8	6	6	6	6	6	6	6	6	6	6	5
DF.s	32	32	32	32	32	32	32	32	32	25	25	25	25	25	25	25	25	25	25	44	43	44	44	43	44	43	43	43	44	3
ID.m	4	4	4	4	4	4	4	4	4	4	4	4	4	4	4	4	4	4	4	3	3	3	3	3	3	3	3	3	3	2
ID.s	51	51	51	51	51	51⌟	51	51	51	17	17	17	17	17	17	17	17	17	17	25	25	25	25	25	25	25	25	25	25	34

⌟ f. 13ᵛ B

⌜c f. 6ᵛ Kh, ⌜d f. 14ᵛ B

	61	62	63	64	65	66	67	⌜c68	69	70	71	72	73	74	75	76	77	78	79	80	81	82	83	⌜d84	85	86	87	88	89	90
EC.d	9	9	9	10	10	10	10	10	10	10	10	10	10	10	10	10	10	10	10	11	11	11	11	11	11	11	11	11	11	11
EC.m	48	53	58	3	8	13	18	23	28	33	36	39	42	44	47	50	53	56	58	1	2	3	5	6	7	8	9	10	11	12
EC.s	13	16	19	21	24	27	30	33	35	38	26	15	3	51	39	28	16	4	52	40	48	55	2	10	17	24	31	39	46	53
DF.m	5	5	5	5	5	5	5	5	5	2	2	2	2	2	2	2	2	2	2	1	1	1	1	1	1	1	1	1	1	0
DF.s	3	3	2	3	3	3	2	3	3	48	49	48	48	48	49	48	48	48	48	8	7	7	8	7	7	7	7	7	7	0
ID.m	2	2	2	2	2	2	2	2	2	1	1	1	1	1	1	1	1	1	1	0	0	0	0	0	0	0	0	0	0	0
ID.s	34	34	34	34	34	34	34⌟e	34	34	25	25	25	25	25	25	25	25	25	25	34	34	34	34⌟f	34	34	34	34	34	34	34⌟g

⌟e f. 6ʳ Kh, ⌟f f. 14ʳ B, ⌟g f. 14ᵛ B, f. 6ᵛ Kh, f. 3ʳ S43, f. 8ʳ S45

Table XII: apparatus criticus

‖ bhaumamandaphalāni ‖] ‖ atha bhaumamandaphalaṃ mandocca 4 | 8 | 30 | 0 ‖ para 81 ‖ (f. 13ᵛ) B; atha bhaumamandaphalādho gatiphalaṃ (f. 6ʳ) Kh, iti bhaumamandaphalam (f. 6ᵛ) Kh; bhaumasya mandaphalāny aṃśādīni ‖ svadeśīmadhyamabhaumanespaṣṭamandoccamāṃthīkādhī itemṛdukendrathā itehatābhujāṃśa uparikoṣṭakejoi (f. 2ᵛ) S43, bhaumasya mandaphalāni ‖ (f. 3ʳ) S43.

EC.s(2) 23 Kh DF.s(4) 46 S43 EC.m(5) 48 B; EC.s(5) 22 S43 EC.s(6) 38 S45 EC.m(9) 46 Kh, 44 S45; EC. ID.s(10) 25 Kh ID.s(11) 25 Kh EC.s(12) 6 B; ID.s(12) 25 Kh ID.s(13) 25 Kh ID.s(14) 25 Kh ID.s(15) 25 Kh ID.s(16) 25 Kh ID.s(17) 25 Kh ID.s(18) 25 Kh ID.s(19) 25 Kh EC.s(22) 7 B EC.s(23) 13 B EC.s(24) 33 B EC.s(30) 28 B EC.s(32) 32 B EC.s(33) 3 B EC.s(34) 36 B DF.s(41) 24 S43 EC.s(42) 32 S43 EC.m(43) 33 S45 DF.s(44) 24 S43 EC.m(45) 56 B; EC.s(45) 57 B EC.m(46) 7 B EC.m(47) 10 B; EC.s(47) 20 B; DF.s(47) 24 S43 EC.s(49) 35 Kh ID.m(50) 4 Kh EC.s(51) 41 B EC.(52) 2 Kh EC.m(54) 20 Kh DF.s(56) 44 S43 EC.s(57) 49 Kh; DF.s(57) 44 S43 EC.s(59) 24 B; DF.s(59) 43 S43 DF.m(60) 6 S43; DF.s(60) 44 S43 DF.s(63) 3 S43 EC.s(64) 11 Kh EC.s(65) 25 B EC.s(66) 26 B DF.s(67) 3 S43, 9 S45 EC.s(68) 23 B; DF.s(68) 2 S43 EC.s(69) 25 S43 EC.d(70) 4 Kh; EC.s(74) 52 B EC.s(76) 25 B EC.m(81) 3 Kh; EC.s(81) 45 B EC.m(82) 5 Kh; EC.s(82) 2 Kh EC.m(83) 6 Kh; EC.s(83) 10 Kh EC.m(84) 7 Kh; EC.s(84) 17 Kh EC.m(85) 8 Kh; EC.s(85) 24 Kh EC.m(86) 9 Kh; EC.s(86) 33 Kh EC.m(87) 10 Kh; EC.s(87) 39 Kh; DF.s(87) 8 S43 EC.m(88) 11 Kh; EC.s(88) 46 Kh EC.m(89) 49 B, 12 Kh; EC.s(89) 53 Kh EC.m(90) 6 Kh; EC.s(90) 21 Kh; DF.m(90) 1 S45; DF.s(90) 7 S45 ID.s(90) 0 S45.

DF.#(1–90) *om.* B • DF.#(1–90) *om.* Kh; EC-table has row labels: "'ṃśa" for EC.d, "ka· viṃ·" for EC.m–EC.s, "gatiḥ" for ID.m–ID.s on f. 6ʳ Kh; EC.s(73) ±3 correction on f. 6ᵛ Kh; EC.#(82)–EC.#(89) values appear left-shifted for commonly attested values corresponding to arguments 83–90 on f. 6ᵛ Kh; EC.m,s(90) 6,21 in Kh appear to be erroneously copied from the values of EC.m,s(1) of Table XIII immediately continued on f. 6ᵛ Kh • "mandakendrabhujāṃśopari antare kalādinā aṅguṇyayojyaṃ sphuṭaṃ phalakendravaśāvarddhanarṇaṃ meṣādau kendre dhanaṃ tulādau ṛṇaṃ adho gatiphalaṃ kendravaśāddhanarṇaṃ karkādau dhanaṃ makarādau ṛṇaṃ ‖" vernacular paratext in the right margin of f. 8ʳ S45 • Table XII *om.* SMB.

Table XIII

Sources		1	2	3	4	5	6	7	8	9	10	11	12	13	14	15	16	17	18	19	20	21	22	23	24	25	26	27	28	29	30		
⌜ f. 16^v B f. 6^v Kh f. 3^r S43 f. 10^r S45	⌜‖ budhamandaphalāni ‖																																
		0	0	0	0	0	0	0	0	0	1	1	1	1	1	1	1	1	1	1	2	2	2	2	2	2	2	2	2	2	3		EC.d
		6	12	19	25	31	38	44	50	57	3	9	15	21	27	33	40	46	52	58	4	10	15	21	27	33	38	44	50	56	1		EC.m
		21	43	5	27	40	10	32	54	16	38	41	45	49	52	56	0	3	7	10	14	0	45	30	16	1	47	32	18	3	49		EC.s
		6	6	6	6	6	6	6	6	6	6	6	6	6	6	6	6	6	6	5	5	5	5	5	5	5	5	5	5	5	5		DF.m
		22	22	22	22	21	22	22	22	22	3	4	4	4	4	4	3	4	3	4	45	45	46	45	46	45	45	46	45	46	9		DF.s
		6	6	6	6	6	6	6	6	6	5	5	5	5	5	5	5	5	5	5	5	5	5	5	5	5	5	5	5	5	4		ID.m
		0	0	0	0	0	0	0	0	0	43	43	43	43	43	43	43	43	43	43	25	25	25	25	25	25	25	25	25	25	51⌟	⌟ f. 3^r S43	ID.s
⌜a f. 3^v S43 ⌜b f. 17^r B		⌜a31	32	33	34	35	⌜b36	37	38	39	40	41	42	43	44	45	46	47	48	49	50	51	52	53	54	55	56	57	58	59	60		
		3	3	3	3	3	3	3	3	3	3	3	4	4	4	4	4	4	4	4	4	4	4	4	4	4	5	5	5	5	5		EC.d
		6	12	17	22	27	32	37	43	48	53	57	2	6	11	16	20	25	29	34	38	42	46	49	53	56	0	4	7	11	15		EC.m
		58	7	16	25	34	43	52	1	10	20	52	25	48	30	3	36	9	41	14	47	25	3	41	20	58	36	14	52	30	9		EC.s
		5	5	5	5	5	5	5	5	5	4	4	4	4	4	4	4	4	4	4	3	3	3	3	3	3	3	3	3	3	2		DF.m
		9	9	9	9	9	9	9	9	10	33	33	33	32	33	33	33	32	33	33	38	39	39	39	38	38	38	38	38	39	43		DF.s
		4	4	4	4	4	4	4	4	4	4	4	4	4	4	4	4	4	4	4	3	3	3	3	3	3	3	3	3	3	2		ID.m
		51	51	51	51	51⌟	51	51	51	51	17	17	17	17	17	17	17	17	17	17	25	25	25	25	25	25	25	25	25	25	34	⌟ f. 16^v B	ID.s
⌜c f. 7^r Kh ⌜d f. 17^v B		61	62	63	64	65	66	67	68	69	⌜c70	71	72	73	74	75	76	77	78	79	⌜d80	81	82	83	84	85	86	87	88	89	90		
		5	5	5	5	5	5	5	5	5	5	5	5	5	5	5	5	5	5	5	5	5	5	5	6	6	6	6	6	6	6		EC.d
		17	20	23	26	28	31	34	36	39	42	43	45	46	48	50	51	53	54	56	57	58	58	59	0	0	1	1	2	3	3	⌟e f. 6^v Kh	EC.m
		52	36	20	3	47	30	14	58	41	25	56	27	58	29	0	30	1	32	3	34	0	47	23	0	36	12	49	25	1	38	⌟f f. 17^r B	EC.s
		2	2	2	2	2	2	2	2	2	1	1	1	1	1	1	1	1	1	1	0	0	0	0	0	0	0	0	0	0	0	⌟g f. 17^v B	DF.m
		44	44	43	44	44	44	44	43	35	31	31	31	31	31	31	31	31	31	31	36	37	36	34	36	36	37	36	36	37	0	f. 7^r Kh	DF.s
		2	2	2	2	2	2	2	2	2	1	1	1	1	1	1	1	1	1	1	0	0	0	0	0	0	0	0	0	0	0	f. 3^v S43	ID.m
		34	34	34	34	34	34	34	34	34	25⌟e	25	25	25	25	25	25	25	25	25⌟f	34	34	34	34	34	34	34	34	34	34	34⌟g	f. 10^r S45	ID.s

Table XIII: apparatus criticus

‖ budhamandaphalāni ‖] ‖ atha budhamandaphalaḥ budhamandaphalaṃ budhamandoccaḥ 7 | 15 | 0 | 0 | para 44 ‖ (f. 16v) B; atha budhamandaphalādho gatiphalaṃ (f. 6v) Kh, iti budhamandaphalam (f. 7r) Kh; atha budhasya mandaphalāny aṃśādīni ‖ (f. 3r) S43, budhasya mandaphalāni ‖ (f. 3v) S43.

EC.m(1) 12 Kh; EC.s(1) 43 Kh, 22 S43 EC.m(2) 19 Kh; EC.s(2) 5 Kh, 44 S43 EC.m(3) 25 Kh; EC.s(3) 27 Kh EC.m(4) 31 Kh; EC.s(4) 49 Kh EC.m(5) 38 Kh; EC.s(5) 10 Kh, 49 S43 EC.m(6) 44 Kh; EC.s(6) 32 Kh EC.m(7) 50 Kh; EC.s(7) 54 Kh EC.s(9) 56 S43 ID.m(10) 6 Kh; ID.s(10) 42 B, 0 Kh ID.s(11) 42 B ID.s(12) 42 B EC.s(13) 30 B; ID.s(13) 42 B EC.s(14) 53 S43; ID.s(14) 42 B ID.s(15) 42 B ID.s(16) 42 B EC.s(17) 20 B; ID.s(17) 42 B EC.m(18) 57 B; ID.s(18) 42 B EC.s(19) 11 S43; ID.s(19) 42 B EC.s(20) 15 S43; DF.s(20) 46 S45; ID.s(20) 43 Kh EC.m(21) 9 Kh; EC.s(21) 59 Kh; ID.s(21) 43 Kh ID.s(22) 43 Kh EC.s(23) 31 S43; ID.s(23) 43 Kh EC.m(24) 28 S43; EC.s(24) 15 S45; DF.s(24) 45 S43; ID.s(24) 43 Kh ID.s(25) 43 Kh EC.s(26) 40 B; ID.s(26) 43 Kh ID.s(27) 43 Kh ID.s(28) 43 Kh DF.s(29) 45 S43; ID.s(29) 43 Kh EC.s(36) 4 Kh EC.s(40) 28 Kh, 2 S45; DF.s(40) 32 S43 EC.s(43) 58 Kh, S45; DF.s(43) 33 S43 EC.s(44) 31 S43 EC.s(45) 4 S43 DF.s(47) 33 S43 EC.s(48) 11 Kh, 42 S43 EC.s(49) 15 S43; DF.s(49) 32 S43 EC.s(50) 57 B; ID.m(50) 4 B, Kh; ID.s(50) 17 B, Kh, 15 S45 EC.s(52) 4 S43; DF.s(52) 38 S43 EC.s(53) 42 S43; DF.s(53) 38 S43 DF.s(54) 38 S43 EC.d(56) 4 Kh; EC.m(56) 1 Kh EC.s(60) 52 B EC.s(64) 83 Kh EC.s(66) 31 S43; DF.s(66) 43 S43 EC.s(67) 44 Kh DF.s(68) 44 S43 EC.s(69) 42 S43; DF.s(69) 43 S43 EC.m(70) 25 B EC.d(73) 4 S45; EC.s(73) 48 B EC.s(76) 31 S43 EC.s(77) 2 S43 EC.s(78) 33 S43 EC.s(79) 4 S43 EC.s(80) 35 Kh, S43 EC.s(81) 10 B, Kh, 11 S43; DF.s(81) 36 S43 EC.s(82) 34 B, 46 Kh DF.s(83) 36 S43 EC.s(86) 52 B; DF.s(86) 36 S43 EC.m(88) 25 B.

DF.#(1–90) *om.* B • DF.#(1–90) *om.* Kh; EC.#(1)–EC.#(7) values appear left-shifted for commonly attested values corresponding to arguments 2–8 on f. 6v Kh; the commonly attested values of EC.m,s(1) 6,21 appear to be erroneously copied to EC.m,s(90) of Table XII immediately preceeded on f. 6v Kh; ID.s(20)–ID.s(29) appear to be erroneously copied as 43 (instead of the commonly attested 25) on f. 6v Kh • Table XIII *om.* SMB.

Table XIV

⌜a f. 19v B, f. 7r Kh, f. 4r S43, f. 12r S45; ⌜b f. 20r B

⌜a‖ gurumandaphalāni ‖

	1	2	3	4	5	6	7	8	9	10	11	12	13	14	15	16	17	18	19	20	21	⌜b22	23	24	25	26	27	28	29	30
EC.d	0	0	0	0	0	0	0	0	0	0	1	1	1	1	1	1	1	1	1	1	1	1	2	2	2	2	2	2	2	2
EC.m	5	11	16	22	27	33	38	44	49	55	0	5	11	16	21	26	32	37	42	47	52	57	2	7	12	17	22	27	32	37
EC.s	31	3	34	6	37	9	41	12	44	15	31	47	3	18	34	50	6	22	37	53	53	53	53	53	53	53	53	53	53	53
DF.m	5	5	5	5	5	5	5	5	5	5	5	5	5	5	5	5	5	5	5	5	5	5	5	5	5	5	5	5	5	4
DF.s	32	31	32	31	32	32	31	32	31	16	16	16	16	16	16	16	16	16	16	0	0	0	0	0	0	0	0	0	0	29
ID.m	0	0	0	0	0	0	0	0	0	0	0	0	0	0	0	0	0	0	0	0	0	0	0	0	0	0	0	0	0	0
ID.s	25	25	25	25	25	25	25	25	25	24	24	24	24	24	24	24	24	24	24	23	23⌟	23	23	23	23	23	23	23	23	20

⌟ f. 19v B

	31	32	33	34	35	36	37	38	39	40	41	42	43	44	45	46	47	48	49	50	51	52	53	54	55	56	57	58	59	60
EC.d	2	2	2	2	3	3	3	3	3	3	3	3	3	3	3	3	3	3	3	4	4	4	4	4	4	4	4	4	4	4
EC.m	42	46	51	55	0	4	9	13	18	22	26	30	34	38	42	46	50	54	58	2	5	8	11	14	17	21	24	27	30	33
EC.s	22	50	18	47	15	43	12	41	9	38	34	31	28	25	22	19	15	12	9	6	15	24	34	44	53	3	12	22	31	41
DF.m	4	4	4	4	4	4	4	4	4	3	3	3	3	3	3	3	3	3	3	3	3	3	3	3	3	3	3	3	3	2
DF.s	28	28	29	28	28	29	29	28	29	56	57	57	57	57	57	57	57	57	57	9	10	9	10	9	10	9	10	9	10	22
ID.m	0	0	0	0	0	0	0	0	0	0	0	0	0	0	0	0	0	0	0	0	0	0	0	0	0	0	0	0	0	0
ID.s	20	20	20	20	20	20	20	20	20	18	18	18	18	18	18	18	18	18	18	14	14	14	14	14	14	14	14	14	14	10⌟

⌟ f. 4r S43

⌜c f. 4v S43; ⌜d f. 20v B; ⌜e f. 7v Kh

	⌜c61	62	63	64	⌜d65	66	67	68	69	70	71	72	73	74	75	76	77	78	79	80	81	82	83	84	85	86	87	⌜e88	89	90
EC.d	4	4	4	4	4	4	4	4	4	4	4	5	5	5	5	5	5	5	5	5	5	5	5	5	5	5	5	5	5	5
EC.m	36	38	40	43	45	47	50	52	55	57	58	0	1	2	3	5	6	7	9	10	11	11	12	12	13	13	14	14	15	15
EC.s	3	25	47	9	31	53	15	37	0	22	41	0	19	37	56	15	34	53	12	31	3	34	6	38	9	41	12	44	15	47
DF.m	2	2	2	2	2	2	2	2	2	1	1	1	1	1	1	1	1	1	1	0	0	0	0	0	0	0	0	0	0	0
DF.s	22	22	22	22	22	22	22	22	22	19	19	19	19	19	19	19	19	19	19	32	31	32	32	31	32	31	32	31	32	0
ID.m	0	0	0	0	0	0	0	0	0	0	0	0	0	0	0	0	0	0	0	0	0	0	0	0	0	0	0	0	0	0
ID.s	10	10	10	10⌟f	10	10	10	10	10	6	6	6	6	6	6	6	6	6	6	2	2	2	2	2	2	2	2⌟g	2	2	2⌟h

⌟f f. 20r B; ⌟g f. 7r Kh; ⌟h f. 20v B, f. 7v Kh, f. 4v S43, f. 12r S45

Table XIV: apparatus criticus

‖ gurumandaphalāni ‖] ‖ atha gurumandaphalaṃ ‖ mandocca 5 | 22 | 30 | 0 | para 23 ‖ (f. 19^{v}) B, gurumandaphalasmāptā ‖ (f. 20^{v}) B; atha gurumandaphalādho gatiphalam (f. 7^{r}) Kh, iti gurumandaphalam (f. 7^{v}) Kh; guror mandaphalāny aṃśādīni ‖ (f. 4^{r}) S43, guror mandaphalāni ‖ (f. 4^{v}) S43;

EC.m(1) 6 B; EC.s(1) 50 B EC.m(2) 1 S45 EC.m(3) 26 S45; EC.s(3)24 B EC.s(5) 28 B EC.m(6) 32 B; EC.s(6) 11 B EC.m(7) 39 B; EC.s(7) 2 B EC.m(8) 45 B; EC.s(8) 11 Kh EC.s(9) 46 B ID.s(10) 25 B EC.s(11) 21 B EC.s(13) 23 B; DF.s(13) 15 S43 EC.m(14) 13 B; EC.s(14) 19 B, 8 Kh EC.s(15) 16 B; ID.s(15) 34 Kh EC.m(16) 25 B EC.s(17) 27 B EC.s(18) 32 B; DF.s(18) 15 S43 EC.s(19) 38 B, Kh, 52 S45 EC.s(20) 52 Kh; ID.s(20) 24 B ID.s(21) 24 B EC.s(23) 43 B EC.m(26) 16 Kh EC.s(29) 33 B ID.s(30) 21 Kh EC.m(31) 43 B; EC.s(31) 4 B; ID.s(31) 21 Kh EC.s(32) 0 B; ID.s(32) 21 Kh DF.m(33) 28 S45; ID.s(33) 21 Kh ID.s(34) 21 Kh EC.d(35) 2 B; EC.m(35) 55 B; EC.s(35) 47 B; ID.s(35) 21 Kh EC.m(36) 0 B; EC.s(36) 15 B, 44 S45; ID.s(36) 21 Kh EC.m(37) 4 B; EC.s(37) 44 B; DF.s(37) 28 S45; ID.s(37) 21 Kh EC.m(38) 9 B; EC.s(38) 12 B, 40 Kh; ID.s(38) 21 Kh EC.m(39) 14 B; EC.s(39) 47 B; ID.s(39) 21 Kh ID.s(40) 45 Kh, 20 S43 EC.m(41) 24 B; EC.s(41) 25 B; ID.s(41) 45 Kh EC.s(42) 21 B; ID.s(42) 45 Kh EC.s(43) 38 B; ID.s(43) 45 Kh EC.s(44) 35 B; ID.s(44) 45 Kh ID.s(45) 45 Kh EC.s(46) 45 B, 18 S45; ID.s(46) 45 Kh EC.m(47) 52 B; EC.s(47) 22 B, 16 S43; ID.s(47) 45 Kh EC.s(48) 52 B ID.s(48) 45 Kh EC.d(49) 4 Kh; ID.s(49) 36 Kh ID.s(50) 36 Kh EC.s(51) 25 B EC.s(52) 25 B, S43; ID.s(52) 36 Kh ID.s(53) 36 Kh EC.m(54) 16 B; EC.s(54) 17 B; ID.s(54) 36 Kh ID.s(55) 36 Kh EC.m(56) 20 B; EC.s(56) 20 B; ID.s(56) 36 Kh DF.s(57) 9 S45; ID.s(57) 36 Kh ID.s(58) 36 Kh ID.s(59) 36 Kh EC.s(60) 43 Kh; ID.s(60) 14 B, 27 Kh ID.s(61) 14 B, 27 Kh EC.s(62) 24 B; ID.s(62) 14 B, 27 Kh ID.s(63) 14 B, 27 Kh EC.m(64) 40 Kh; ID.s(64) 14 B, 27 Kh EC.m(65) 41 Kh; EC.s(65) 27 B; ID.s(65) 27 Kh EC.m(66) 27 B; ID.s(66) 27 Kh EC.s(67) 3 S45; ID.s(67) 27 Kh EC.s(68) 25 B; ID.s(68) 27 Kh ID.s(69) 27 Kh ID.s(70) 16 Kh ID.s(71) 16 Kh ID.s(72) 16 Kh EC.m(73) 0 B; EC.s(73) 29 B, 9 S43; ID.s(73) 16 Kh ID.s(74) 16 Kh EC.s(75) 36 Kh; ID.s(75) 16 Kh ID.s(76) 16 Kh EC.s(77) 35 Kh; ID.s(77) 16 Kh ID.s(78) 16 Kh EC.s(79) 10 B, 13 Kh DF.s(79) 32 S43; ID.s(79) 16 Kh EC.s(80) 30 Kh; ID.s(80) 6 Kh DF.s(81) 32 S43; ID.s(81) 6 Kh DF.s(81) 32 S43; ID.s(82) 6 Kh ID.s(83) 6 Kh EC.s(84) 39 B, 36 Kh; DF.s(84) 32 S43; ID.s(84) 6 Kh ID.s(85) 6 Kh EC.m(86) 14 B; EC.s(86) 12 B; DF.s(86) 3 S43; ID.s(86) 6 Kh EC.s(87) 14 Kh; ID.s(87) 6 Kh EC.d(88) 4 Kh; DF.s(88) 32 S43; ID.s(88) 6 Kh EC.d(89) 4 Kh; ID.s(89) 6 Kh EC.d(90) 4 Kh; DF.s(90) 31 S45; ID.s(90) 6 Kh.

DF.#(1–90) *om.* B; EC.#(34) and EC.#(35) repeated in B; EC.#(36)–EC.#(39) values appear right-shifted for commonly attested values corresponding to arguments 35–38 on f. 20^{r} B • DF.#(1–90) *om.* Kh; ID.m s(51) *om.* Kh; the values: ID.s(30)–ID.s(39) are 21, ID.s(40)–ID.s(48) are 45, ID.s(49)–ID.s(59) are 36, ID.s(60)–ID.s(69) are 27, ID.s(70)–ID.s(79) are 16, ID.s(80)–ID.s(90) are 6 on f. 7rv Kh • EC-table has row labels: "'ṃśa" for arguments: 61–90, "'ṃśa" for EC.d(61–90), "kalā" for EC.m(61–90), "vikalā" for EC.s(61–90), "'ṃtara" for DF.m–DF.s(61–90), "gatiphala" for ID.m–ID.s(61–90) in the right margin of f. 4^{v} S43 • Table XIV *om.* SMB.

Table XV

⌜a f. 22^{v} B
f. 7^{v} Kh
f. 4^{v} S43
f. 14^{r} S45
⌜b f. 23^{r} B

⌜a‖ śukramandaphalāni ‖

	1	2	3	⌜b4	5	6	7	8	9	10	11	12	13	14	15	16	17	18	19	20	21	22	23	24	25	26	27	28	29	30
EC.d	0	0	0	0	0	0	0	0	0	0	0	0	0	0	0	0	0	0	0	0	0	0	0	0	0	0	0	0	0	0
EC.m	1	3	4	6	8	9	11	12	14	16	17	19	20	22	23	25	26	28	29	31	32	34	35	37	38	40	41	43	44	45
EC.s	36	12	49	25	2	38	15	51	27	4	36	7	39	11	43	15	47	19	50	22	49	17	44	11	38	6	34	1	28	55
DF.m	1	1	1	1	1	1	1	1	1	1	1	1	1	1	1	1	1	1	1	1	1	1	1	1	1	1	1	1	1	1
DF.s	36	37	36	37	36	37	36	36	36	32	31	32	32	32	32	32	32	31	32	27	28	27	27	27	28	28	27	27	27	18
ID.m	1	1	1	1	1	1	1	1	1	1	1	1	1	1	1	1	1	1	1	1	1	1	1	1	1	1	1	1	1	1
ID.s	45	45	45⌟e	45	45	45	45	45	45	40	40	40	40	40	40	40	40	40	40	35	35	35	35	35	35	35	35	35	35	25⌟f

⌟e f. 22^{v} B
⌟f f. 4^{v} S43

⌜c f. 5^{r} S43
⌜d f. 23^{v} B

	⌜c31	32	33	34	35	36	37	38	39	40	41	42	43	44	45	46	47	48	49	50	51	⌜d52	53	54	55	56	57	58	59	60
EC.d	0	0	0	0	0	0	0	0	0	0	1	1	1	1	1	1	1	1	1	1	1	1	1	1	1	1	1	1	1	1
EC.m	47	48	49	51	52	53	55	56	57	58	0	1	2	3	4	5	6	8	9	10	11	12	13	14	15	15	16	17	18	19
EC.s	13	31	49	7	25	43	1	19	37	55	4	13	22	31	40	49	57	6	15	24	19	14	9	4	0	55	50	45	40	35
DF.m	1	1	1	1	1	1	1	1	1	1	1	1	1	1	1	1	1	1	1	0	0	0	0	0	0	0	0	0	0	0
DF.s	18	18	18	18	18	18	18	18	18	9	9	9	9	9	9	9	9	9	9	55	55	55	55	55	55	55	55	55	55	41
ID.m	1	1	1	1	1	1	1	1	1	1	1	1	1	1	1	1	1	1	1	1	1	1	1	1	1	1	1	1	1	0
ID.s	25	25	25	25	25	25	25	25	25	15	15	15	15	15	15	15	15	15	15	0	0⌟	0	0	0	0	0	0	0	0	45

⌟ f. 23^{r} B

	61	62	63	64	65	66	67	68	69	70	71	72	73	74	75	76	77	78	79	80	81	82	83	84	85	86	87	88	89	90
EC.d	1	1	1	1	1	1	1	1	1	1	1	1	1	1	1	1	1	1	1	1	1	1	1	1	1	1	1	1	1	1
EC.m	20	20	21	22	23	23	24	25	25	26	26	27	27	28	28	28	29	29	29	30	30	30	30	30	31	31	31	31	31	31
EC.s	16	58	39	20	2	43	24	6	47	28	51	14	37	0	23	46	9	32	55	18	27	36	45	55	4	13	22	31	41	50
DF.m	0	0	0	0	0	0	0	0	0	0	0	0	0	0	0	0	0	0	0	0	0	0	0	0	0	0	0	0	0	0
DF.s	42	41	41	42	41	41	42	41	41	23	23	23	23	23	23	23	23	23	23	9	9	9	9	10	9	9	9	9	9	0
ID.m	0	0	0	0	0	0	0	0	0	0	0	0	0	0	0	0	0	0	0	0	0	0	0	0	0	0	0	0	0	0
ID.s	45	45	45	45	45	45	45	45	45	25	25	25	25	25	25	25	25	25	25	10	10	10	10	10	10	10	10	10	10	10⌟

⌟ f. 23^{v} B
f. 7^{v} Kh
f. 5^{r} S43
f. 14^{r} S45

Table XV: apparatus criticus

‖ śukramandaphalāni ‖] ‖ śu ∘ manda ∘ phalam ‖ (f. 22v) B, śukramandaphalaṃ ‖ mandocca 2 | 21 | 0 | 0 | para 87 ‖ (f. 23r) B, śukramandaphalasamāpatāḥ ‖ (f. 23v) B; bhṛgumandaphalādho gatiphalam iti bhṛgumandaphalam (f. 7v) Kh; atha śukrasya mandaphalāny aṃśādīni ‖ (f. 4v) S43, bhṛgor mandaphalāni ‖ (f. 5r) S43.

EC.s(3) 29 B, 48 Kh DF.s(5) 37 S45 ID.s(7) 4 Kh EC.s(8) 40 B; ID.s(8) 4 Kh DF.s(9) 37 S43; ID.s(9) 4 Kh EC.s(10) 40 B EC.m(12) 18 B; EC.s(12) 34 B EC.m(14) 23 B; EC.s(14) 43 B EC.m(15) 24 B; EC.s(15) 39 B EC.m(18) 26 B; EC.s(18) 47 B EC.m(19) 19 B DF.s(19) 28 S45 EC.s(20) 32 Kh, 20 S43 DF.s(21) 27 S45 EC.m(27) 43 B; EC.s(27) 0 B EC.m(28) 44 B; EC.s(28) 27 B EC.m(29) 45 B; EC.s(29) 55 B DF.s(30) 28 S45 EC.s(32) 21 B, 30 Kh EC.m(34) 50 B EC.s(40) 8 S43 EC.s(43) 23 B EC.s(46) 45 B EC.m(48) 7 Kh EC.m(49) 8 Kh EC.m(50) 9 Kh; EC.m(50) 14 Kh; DF.s(50) 55 S43 EC.m(51) 10 Kh EC.m(52) 11 Kh EC.m(53) 12 Kh EC.m(54) 13 Kh EC.m(55) 14 Kh EC.s(59) 41 Kh EC.s(61) 6 B EC.m(62) 21 Kh; DF.s(62) 42 S45 EC.s(65) 3 Kh EC.s(66) 44 Kh EC.s(70) 18 B, 25 Kh EC.s(73) 36 B EC.s(75) 20 S45 EC.s(76) 26 Kh EC.s(78) 55 B EC.s(79) 58 B ID.s(80) 9 Kh ID.s(81) 9 Kh EC.s(82) 36 B; ID.s(82) 9 Kh ID.s(83) 9 Kh DF.s(84) 9 S43; ID.s(84) 9 Kh ID.s(85) 9 Kh ID.s(86) 9 Kh EC.s(87) 32 B; ID.s(87) 9 Kh EC.s(88) 36 B; ID.s(89) 9 Kh EC.s(89) 40 S43; DF.s(89) 10 S43; ID.s(89) 9 Kh ID.s(90) 0 S43.

DF.#(1–90) *om.* B; ID.m(3) and ID.s(3) corrected to $\frac{1}{45}$ in the lower margin on f. 22v B • DF.#(1–90) *om.* Kh; ID.s(90) *om.* f. 7v Kh; values ID.s(80)–ID.s(90) are 9 on f. 7v Kh • EC-table has row labels: "'ṃśa" for arguments: 1–30, "'ṃśa" for EC.d(1–30), "kalā" for EC.m(1–30), "vika" for EC.s(1–30), "'ṃtara" for DF.m–DF.s(1–30), "gati" for ID.m–ID.s(1–30) in the right margin of f. 4v S43 • Table XV *om.* SMB.

Table XVI

⌜ f. 26r B
f. 8r Kh
f. 5r S43
f. 16r S45

	⌜‖ śanimandaphalāni ‖																													
	1	2	3	4	5	6	7	8	9	10	11	12	13	14	15	16	17	18	19	20	21	22	23	24	25	26	27	28	29	30
EC.d	0	0	0	0	0	0	0	1	1	1	1	1	1	1	1	2	2	2	2	2	2	2	2	3	3	3	3	3	3	3
EC.m	8	16	24	32	40	48	56	4	12	20	27	35	43	50	58	6	13	21	29	36	43	51	58	5	12	20	27	34	42	49
EC.s	1	3	4	6	7	9	10	12	13	15	53	32	11	49	28	6	45	24	2	41	56	12	28	43	59	15	31	46	2	18
DF.m	8	8	8	8	8	8	8	8	8	7	7	7	7	7	7	7	7	7	7	7	7	7	7	7	7	7	7	7	7	6
DF.s	2	1	2	1	2	1	2	1	2	38	39	39	38	39	38	39	39	38	39	15	16	16	15	16	16	16	15	16	16	29
ID.m	0	0	0	0	0	0	0	0	0	0	0	0	0	0	0	0	0	0	0	0	0	0	0	0	0	0	0	0	0	0
ID.s	10	10	10	10	10	10	10	10	10	10	10	10	10	10	10	10	10	10	10	9	9	9	9	9	9	9	9	9	9	8⌟

⌟ f. 5r S43

⌜a f. 5v S43
⌜b f. 26v B

	⌜a31	32	33	34	35	36	37	38	39	40	41	42	43	44	45	46	⌜b47	48	49	50	51	52	53	54	55	56	57	58	59	60
EC.d	3	4	4	4	4	4	4	4	4	4	5	5	5	5	5	5	5	5	5	5	5	6	6	6	6	6	6	6	6	6
EC.m	55	2	8	15	21	28	34	41	47	54	0	5	11	17	22	28	34	40	45	51	56	0	5	9	14	19	23	28	32	37
EC.s	47	17	47	17	47	16	46	16	46	16	0	43	27	11	55	39	23	7	51	35	10	45	21	56	31	6	41	16	51	27
DF.m	6	6	6	6	6	6	6	6	6	5	5	5	5	5	5	5	5	5	5	4	4	4	4	4	4	4	4	4	4	3
DF.s	30	30	30	30	30	30	30	30	30	44	43	44	44	44	44	44	44	44	44	35	35	35	35	35	35	35	35	35	36	36
ID.m	0	0	0	0	0	0	0	0	0	0	0	0	0	0	0	0	0	0	0	0	0	0	0	0	0	0	0	0	0	0
ID.s	8	8	8	8	8	8	8	8	8	7	7	7	7	7	7	7⌟	7	7	7	6	6	6	6	6	6	6	6	6	6	4

⌟ f. 26r B

	61	62	63	64	65	66	67	68	69	70	71	72	73	74	75	76	77	78	79	80	81	82	83	84	85	86	87	88	89	90
EC.d	6	6	6	6	6	6	7	7	7	7	7	7	7	7	7	7	7	7	7	7	7	7	7	7	7	7	7	7	7	7
EC.m	40	44	47	51	54	58	1	4	8	11	13	15	17	19	21	23	25	27	29	30	31	32	33	34	34	35	36	37	37	38
EC.s	53	19	46	12	39	5	31	58	24	50	45	40	34	29	24	18	13	8	2	57	43	29	14	0	46	32	18	4	50	36
DF.m	3	3	3	3	3	3	3	3	3	1	1	1	1	1	1	1	1	1	1	0	0	0	0	0	0	0	0	0	0	0
DF.s	26	27	26	27	26	27	27	26	26	55	55	54	55	55	54	55	55	54	55	46	46	46	46	46	46	46	46	46	46	0
ID.m	0	0	0	0	0	0	0	0	0	0	0	0	0	0	0	0	0	0	0	0	0	0	0	0	0	0	0	0	0	0
ID.s	4	4	4	4	4	4	4	4	4	2	2	2	2	2	2	2	2	2	2	1	1	1	1	1	1	1	1	1	1	0⌟

⌟ f. 26 B
f. 8r Kh
f. 5v S43
f. 16r S45

Table XVI: apparatus criticus

|| śanimandaphalāni ||] || śanimandaphalaṃ mandocca | 7 | 28 | 0 | 0 | para 13 || (f. 26r) B; śanimandaphalādho gatiphalam śanimandaphalasaṃpūrṇam || (f. 8r) Kh; śaner mandaphalāny aṃśādīni || (f. 5r) S43, śaner mandaphalāni || (f. 5v) S43.

EC.s(4) 1 Kh EC.s(5) 8 Kh, 2 S43 EC.s(7) 15 B EC.s(8) 11 Kh EC.s(9) 14 B DF.s(10) 39 S43 EC.m(11) 28 B; EC.m(11) 16 B EC.m(12) 36 B; EC.s(12) 18 B, 33 Kh EC.m(13) 44 B; EC.s(13) 20 B; DF.s(13) 39 S43 EC.m(14) 51 B; EC.s(14) 24 B EC.s(15) 26 B; DF.s(15) 39 S43 EC.s(16) 16 B EC.m(17) 14 Kh EC.m(18) 22 Kh; EC.s(18) 14 B, Kh EC.m(19) 30 Kh; ID.s(19) 9 B EC.m(20) 37 Kh; EC.s(20) 42 S45 EC.s(21) 36 B EC.s(22) 52 B, S45 EC.s(23) 48 B EC.s(24) 44 B EC.s(27) 51 B DF.s(28) 16 S43 DF.s(29) 16 S43 Df.s(30) 30 S43; ID.s(30) 9 B EC.m(31) 57 S43 EC.s(33) 40 B, 45 Kh EC.s(35) 41 Kh EC.d(40) 5 B; EC.m(40) 45 B; EC.s(40) 12 B DF.s(41) 44 S45 EC.s(45) 5 B EC.s(47) 27 B, 33 Kh EC.s(49) 57 B; ID.s(49) 6 B EC.s(50) 31 B EC.s(55) 30 B EC.s(56) 2 B EC.s(57) 42 B, 51 S43 DF.s(59) 35 S45 EC.m(60) 36 B; EC.s(60) 51 B EC.m(62) 43 B; EC.s(62) 49 B, 18 Kh EC.s(63) 56 B EC.s(67) 30 Kh, 32 S43; DF.s(67) 26 S43 EC.s(70) 47 B; ID.m(70) 2 S45; ID.s(70) 0 S45 EC.s(71) 53 B; ID.m(71) 2 S45; ID.s(71) 0 S45 DF.s(72) 55 S45; ID.m(72) 2 S45; ID.s(72) 0 S45 ID.m(73) 2 S45; ID.s(73) 0 S45 EC.s(74) 3 B; ID.m(74) 2 S45; ID.s(74) 0 S45 EC.s(75) 45 B; DF.s(75) 55 S45; ID.m(75) 2 S45; ID.s(75) 0 S45 ID.m(76) 2 S45; ID.s(76) 0 S45 ID.m(77) 2 S45; ID.s(77) 0 S45 DF.s(78) 55 S45; ID.m(78) 2 S45; ID.s(78) 0 S45 ID.m(79) 2 S45; ID.s(79) 0 S45 DF.s(82) 45 S43 EC.s(83) 4 B, 15 Kh EC.s(85) 56 B EC.s(88) 40 B, 41 Kh EC.d(89) 0 S45 EC.s(90) 35 B, S45; DF.s(90) 45 S45; ID.s(90) 0 S45.

DF.#(1–90) *om.* B • DF.#(1–90) *om.* Kh • commonly attested values of ID.m,s(70–79) as 0,2 transposed to 2,0 on f. 16r S45 • Table XVI *om.* SMB.

Table XVII-A

	⌜bhaumaśīghraphalaṃḥ																															
⌜f. 14v B	1	2	3	4	5	6	7	8	9	10	11	12	13	14	15	16	17	18	19	20	21	22	23	24	25	26	27	28	29	30		
	0	0	1	1	2	2	2	3	3	4	4	4	5	5	5	6	6	7	7	7	8	8	9	9	9	10	10	11	11	11		EJ.d
	24	48	12	36	0	25	49	13	37	1	25	48	11	34	58	21	44	8	31	55	18	40	3	26	49	13	37	2	26	51		EJ.m
	11	22	33	43	55	5	17	30	40	52	4	18	34	52	12	32	55	20	49	15	4	42	20	50	47	32	22	19	48	23		EJ.s
	200	200	200	200	200	200	200	199	199	199	199	199	198	198	198	198	197	197	197	197	196	195	195	195	195	194	194	193	193	193		AH.a
	15	44	34	28	20	13	56	56	49	40	26	11	56	41	26	11	45	19	56	11	45	19	56	31	41	41	25	51	21	21		AH.b
⌜f. 15r B	31	32	33	34	35	36	37	38	39	40	41	⌜42	43	44	45	46	47	48	49	50	51	52	53	54	55	56	57	58	59	60		
	12	12	12	13	13	14	14	14	15	15	16	16	16	17	17	17	18	18	19	19	19	20	20	21	21	21	22	22	23	23		EJ.d
	14	36	59	13	45	1	32	55	19	53	4	27	49	12	34	57	27	45	9	32	54	17	42	3	26	49	23	37	1	26		EJ.m
	5	49	44	45	52	19	41	48	21	1	55	3	17	33	9	19	17	52	28	21	43	20	59	5	20	48	36	30	54	14		EJ.s
	192	192	192	190	189	189	189	188	187	187	186	186	185	184	184	183	182	182	181	181	180	179	179	178	177	176	176	175	174	173		AH.a
	28	52	23	50	17	45	32	36	33	33	52⌟	11	30	49	9	36	56	20	44	17	28	43	1	18	35	52	8	25	41	51	⌟f. 14v B	AH.b
⌜f. 15v B	61	62	63	64	65	66	67	68	69	70	71	72	73	74	75	76	77	78	79	80	81	82	83	84	85	86	87	⌜88	89	90		
	23	24	24	24	25	25	26	26	26	27	27	27	28	28	28	29	29	29	30	30	30	31	31	31	32	32	33	33	33	34		EJ.d
	47	2	30	52	15	37	0	23	47	11	48	48	7	27	47	7	27	48	10	34	54	14	25	56	17	39	1	24	48	12		EJ.m
	28	55	44	40	8	40	38	52	18	13	34	34	45	11	3	9	47	32	46	13	31	40	20	18	35	45	54	28	2	34		EJ.s
	172	171	171	170	169	168	168	167	166	165	164	162	162	161	160	159	158	157	156	155	154	153	152	151	150	149	148	147	145	144		AH.a
	17	26	26	36	14	52	1	9	17	24	38	32	35	38	40	42	43	44	45	44	42	38	33	28	23	17	10⌟	3	55	46	⌟f. 15r B	AH.b
	91	92	93	94	95	96	97	98	99	100	101	102	103	104	105	106	107	108	109	110	111	112	113	114	115	116	117	118	119	120		
	34	34	34	35	35	35	36	36	36	36	37	37	37	37	37	38	38	38	38	39	39	39	39	39	39	40	40	40	40	40		EJ.d
	26	41	56	12	38	45	2	20	38	56	8	20	33	46	59	13	28	43	59	57	22	39	36	43	51	0	11	22	34	47		EJ.m
	26	15	27	27	25	0	12	2	14	44	29	29	4	14	42	48	24	40	53	53	28	40	1	35	44	38	20	51	31	52		EJ.s
	143	142	141	139	138	137	136	134	133	132	131	129	128	127	125	124	123	122	120	119	118	116	115	114	112	111	109	108	107	105		AH.a
	46	31	8	53	39	24	8	51	34	17	2	47	31	14	57	39	20	1	40	15	2	44	24	4	43	21	58	34	10	44	(continues)	AH.b

Table XVII-A

(continued) ⌜ f. 16ʳ B	121	122	123	124	125	126	127	128	129	130	131	132	133	134	⌜135	136	137	138	139	140	141	142	143	144	145	146	147	148	149	150		
	40	40	40	40	40	40	41	41	41	41	41	41	40	40	40	40	40	40	40	40	40	39	39	39	39	38	38	38	38	38		EJ.d
	48	48	50	25	54	57	1	3	11	18	9	0	51	43	36	34	33	32	32	32	16	59	3	9	13	58	44	39	17	5		EJ.m
	17	46	21	0	48	46	20	52	50	16	15	7	51	36	48	38	14	56	23	8	58	3	13	11	20	40	0	23	9	4		EJ.s
	104	103	101	100	99	97	96	94	93	92	90	89	88	86	85	84	82	81	79	78	76	75	74	72	71	70	69	67	66	64		AH.a
	24	8	48	28	16	44	21	56	30	3	45	27	7	42⌟	24	59	39	9	38	8	54	38	20	20	44	23	1	38	10	41	⌟ f. 15ᵛ B	AH.b
																									bhaumaśīghraphalaṃḥ ‖							

⌜ f. 16ᵛ B	151	152	153	154	155	156	157	158	159	160	161	162	163	164	165	166	167	168	169	170	171	172	173	174	⌜175	176	177	178	179	180		
	37	36	36	35	35	34	33	33	32	31	30	29	28	26	25	24	33	21	20	18	17	15	12	11	9	7	5	4	2	0		EJ.d
	29	53	15	58	19	20	43	3	23	25	28	17	8	56	29	25	8	46	22	58	12	24	34	41	48	55	55	3	3	0		EJ.m
	13	17	53	59	34	57	30	34	37	25	8	50	32	43	5	40	8	47	44	59	57	51	28	49	16	17	17	6	6	0		EJ.s
	63	62	61	60	58	56	56	54	53	52	51	50	49	48	48	47	46	45	44	43	42	42	42	41	41	40	40	39	39	36	⌟a f. 16ʳ B	AH.a
	13	33	13	0	5	13	13	54	21	9	31	32	41	50	5	6	42	17	21	23	58	33	7	41⌟a	15	48	22	56	27	52⌟b	⌟b f. 16ᵛ B	AH.b

Table XVII-B

⌜f. 8v Kh	⌜1	2	3	4	5	6	7	8	9	10	11	12	13	14	15	16	17	18	19	20	21	22	23	24	25	26	27	28	29	30		
	0	0	1	1	2	2	2	3	3	4	4	4	5	5	5	6	6	7	7	7	8	6	9	9	9	10	10	11	11	11		EJ.d
	24	48	12	36	0	25	49	13	37	1	25	48	11	34	58	21	44	8	31	55	18	40	2	26	49	13	37	2	26	51		EJ.m
	10	21	32	43	54	5	17	49	40	52	4	18	34	51	11	31	55	10	47	16	2	51	45	44	46	31	52	18	47	12		EJ.s
	200	200	200	200	200	20	200	199	199	199	199	199	198	198	198	198	197	197	197	197	96	96	95	95	95	94	94	94	94	94		AH.a
	55	44	36	28	21	13	5	5	49	42	27	12	57	41	26	12	56	41	26	11	46	21	56	31	6	41	10	16	51	26		AH.b
	‖ atha bhaumaśīghraphalādhaḥ karṇaḥ ‖																															
⌜f. 9r Kh	31	32	33	34	35	36	37	⌜38	39	40	41	42	43	44	45	46	47	48	49	50	51	52	53	54	55	56	57	58	59	60		
	12	12	12	13	13	14	14	14	15	15	16	16	16	17	17	17	18	18	19	19	19	20	20	21	21	21	22	22	23	23		EJ.d
	14	36	59	22	45	9	32	55	12	42	4	26	49	11	34	57	21	45	9	33	55	17	40	3	26	49	13	37	1	26		EJ.m
	32	49	42	42	40	1	20	46	19	59	55	58	12	34	4	37	22	18	26	47	0	17	2	3	18	47	39	18	42	10		EJ.s
	93	93	92	91	91	19	190	89	88	87	86	85	86	84	84	83	82	82	81	81	80	79	79	78	76	76	76	75	74	73		AH.a
	26	1	29	51	23	52	42⌟	12	33	23	23	53	53	50	9	33	47	3	44	8	26	43	1	8	35	25	25	25	41	57	⌟ f. 8v Kh	AH.b
	61	62	63	64	65	66	67	68	69	70	71	72	73	74	75	76	77	78	79	80	81	82	83	84	85	86	87	88	89	90		
	23	24	24	24	25	25	26	26	26	27	27	27	28	28	28	29	29	22	30	30	30	31	31	31	32	33	33	33	33	34		EJ.d
	47	8	30	52	14	37	0	23	47	11	28	48	7	27	46	7	27	48	11	34	54	14	35	56	17	39	1	24	47	11		EJ.m
	23	54	37	39	46	25	28	42	15	7	48	33	43	11	59	7	34	43	41	45	29	36	8	6	30	33	43	34	58	53		EJ.s
	73	72	71	7	69	68	67	66	65	64	64	63	160	160	160	59	48	57	56	55	54	53	52	51	150	49	48	47	146	45		AH.a
	18	17	27	26	45	54	2	10	17	25	25	28	38	38	40	44	44	45	41	46	42	38	34	29	23	10	11	3	55	43		AH.b
	91	92	93	94	95	96	97	98	99	100	101	102	103	104	105	106	107	108	109	110	111	112	113	114	115	116	117	118	119	120		
	34	34	34	35	35	35	36	36	36	36	37	37	37	37	37	38	38	38	38	39	39	39	39	39	40	40	40	40	40	40		EJ.d
	26	41	56	11	28	45	2	19	37	56	8	22	33	46	59	13	26	43	59	16	22	29	35	43	51	0	11	22	35	42		EJ.m
	21	11	27	27	58	0	5	45	58	41	18	49	1	1	37	42	19	32	31	2	8	50	55	44	44	31	28	31	38	55		EJ.s
	143	42	41	39	38	37	36	34	33	32	31	29	28	27	26	24	23	22	120	119	118	17	16	114	112	110	101	8	17	5		AH.a
	35	11	8	40	39	24	8	52	35	17	3	47	31	15	15	39	31	1	41	2	23	41	24	4	43	21	58	34	1	44		AH.b

(continues)

Table XVII-B

(continued)
⌜ f. 9v Kh

121	122	123	124	125	126	127	128	129	130	131	132	133	134	135	136	137	138	139	140	141	⌜142	143	144	145	146	147	148	149	150		
40	40	40	40	40	41	41	41	41	41	41	40	40	40	40	40	40	40	40	14	40	39	39	39	39	38	38	38	38	37		EJ.d
44	47	49	51	54	57	21	6	17	17	8	0	51	40	36	30	32	32	30	22	11	58	42	20	27	58	43	30	17	4		EJ.m
58	40	46	53	17	31	27	32	52	49	57	5	49	5	48	28	55	6	36	55	7	29	55	35	34	18	51	23	6	53		EJ.s
4	103	101	100	99	97	96	94	93	92	90	89	88	86	85	84	82	81	79	78	76	75	74	73	71	70	69	67	66	64		AH.a
27	8	49	28	7	44	21	56	21	4	46	27	7	46	24	0	35	7	38	8	54⌟	39	22	5	5	24	1	38	10	48	⌟ f. 9r Kh	AH.b

151	152	153	154	155	156	157	158	159	160	161	162	163	164	165	166	167	168	169	170	171	172	173	174	175	176	177	178	179	180		
37	36	36	35	35	34	33	33	32	31	30	29	28	26	25	24	23	21	20	18	17	15	13	11	9	7	6	4	2	0		EJ.d
29	52	15	38	0	22	42	3	22	41	28	17	7	56	42	24	7	46	22	58	12	24	34	41	47	51	7	3	3	0		EJ.m
10	49	54	18	40	10	59	33	55	46	3	50	41	9	30	16	53	50	47	21	55	12	12	39	5	51	27	11	4	0		EJ.s
63	62	61	60	58	57	56	54	53	51	51	50	49	48	47	47	46	45	44	42	42	42	41	40	40	40	39	39	39	39		AH.a
33	34	13	1	46	31	14	54	33	10	31	32	42	51	56	6	12	17	21	58	33	8	41	16	49	34	55	28	28	1⌟	⌟ f. 9v Kh	AH.b

Table XVII-C

⌜‖ bhaumaśīghraphalaṃ ‖

	1	2	3	4	5	6	7	8	9	10	11	12	13	14	15	16	17	18	19	20	21	22	23	24	25	26	27	28	29	30		
⌜ f. 6^{r} S43 f. 8^{v} S45	0	0	1	1	2	2	2	3	3	4	4	4	5	5	5	6	6	7	7	7	8	8	9	9	9	10	10	11	11	11		EJ.d
	24	48	12	36	0	25	49	13	37	1	24	48	11	34	58	21	44	8	31	55	18	40	3	26	49	13	37	2	26	51		EJ.m
	10	21	32	43	54	5	17	28	40	52	4	18	34	51	11	32	55	20	47	16	2	51	45	44	46	31	52	18	47	22		EJ.s
	24	24	24	24	24	24	24	24	24	23	23	23	23	23	23	23	23	23	23	22	22	22	22	23	24	24	24	24	24	23		DF.m
	11	11	11	12	12	12	12	12	12	12	14	16	17	20	21	23	25	27	29	46	49	54	59	2	45	11	26	29	35	10		DF.s
	200	200	200	200	200	200	200	199	199	199	199	199	198	198	198	198	197	197	197	197	196	196	195	195	195	194	194	193	193	193		AH.a
	52	44	26	28	20	13	5	57	49	41	26	11	56	41	36	11	56	41	26	10	45	21	56	31	6	41	16	51	26	1		AH.b
	10	20	31	42	57	7	18	21	32	44	41	40	29	24	21	18	7	1	2	54	57	9	9	16	24	18	22	1	15	16⌟	⌟ f. 6^{r} S43	AH.c
⌜ f. 6^{v} S43	⌜31	32	33	34	35	36	37	38	39	40	41	42	43	44	45	46	47	48	49	50	51	52	53	54	55	56	57	58	59	60		
	12	12	12	13	13	14	14	14	15	15	16	16	16	17	17	17	18	18	19	19	19	20	20	21	21	21	22	22	23	23		EJ.d
	14	46	59	22	45	9	32	55	15	42	4	26	49	11	34	57	21	45	9	33	55	17	40	3	26	49	13	37	1	26		EJ.m
	32	49	43	42	48	1	20	46	19	59	45	58	12	38	4	37	22	18	26	47	0	17	2	3	18	47	29	28	42	10		EJ.s
	22	22	23	23	23	23	23	23	23	21	22	22	22	22	23	23	23	24	24	21	22	22	23	23	23	23	23	24	24	21		DF.m
	17	53	0	6	13	19	26	33	40	46	3	14	22	30	33	45	56	8	11	13	17	45	1	15	29	42	59	14	28	13		DF.s
	192	191	191	190	190	189	189	188	188	187	186	186	185	184	184	183	182	182	181	181	180	179	179	178	177	176	176	175	174	173		AH.a
	28	56	23	50	17	45	12	39	6	33	52	11	30	49	8	32	56	20	44	7	25	43	1	18	35	52	8	25	41	57	⌟ f. 6^{v} S43	AH.b
	38	0	25	40	59	7	20	20	28	21	37	56	58	57	58	58	51	41	13	44	39	23	5	19	21	11	56	16	23	26⌟	f. 8^{v} S45	AH.c
⌜ f. 7^{r} S43 f. 9^{r} S45	⌜61	62	63	64	65	66	67	68	69	70	71	72	73	74	75	76	77	78	79	80	81	82	83	84	85	86	87	88	89	90		
	23	24	24	24	25	25	26	26	26	27	27	27	28	28	28	29	29	29	30	30	30	31	31	31	32	32	33	33	33	34		EJ.d
	47	8	30	52	14	37	0	23	47	11	28	48	7	27	46	7	27	48	10	34	54	14	35	56	17	39	1	24	47	11		EJ.m
	23	52	37	39	46	35	28	42	15	7	48	33	43	11	59	7	34	43	41	45	29	36	8	6	30	23	43	34	58	53		EJ.s
	21	21	22	22	22	22	23	23	23	17	19	19	19	19	20	20	21	21	24	19	20	20	20	21	21	22	21	22	23	24		DF.m
	29	45	2	7	49	53	14	33	52	41	45	10	28	48	8	27	9	58	4	44	7	32	58	24	53	30	51	24	55	28		DF.s
	173	172	171	170	169	168	168	167	166	165	164	163	162	161	160	159	158	157	156	155	154	153	152	151	150	149	148	147	145	144		AH.a
	7	17	26	36	44	53	1	9	17	24	28	32	35	38	40	42	43	44	45	45	42	38	33	28	23	17	10	3	55	46		AH.b
	34	22	52	4	58	32	57	49	22	34	31	6	15	0	21	16	46	50	27	37	12	17	49	50	18	12	33	17	15	46		AH.c

(continues)

Table XVII-C

(continued)

	91	92	93	94	95	96	97	98	99	100	101	102	103	104	105	106	107	108	109	110	111	112	113	114	115	116	117	118	119	120
EJ.d	34	34	34	35	35	35	36	36	36	36	37	37	37	37	37	38	38	38	38	39	39	39	39	39	39	40	40	40	40	40
EJ.m	26	41	56	11	28	45	2	19	37	56	8	22	33	46	59	13	28	43	59	16	22	29	35	43	51	0	11	22	34	47
EJ.s	21	11	27	37	58	0	5	47	58	41	18	49	0	1	37	42	19	32	31	2	8	50	55	28	44	30	28	32	38	45
DF.m	14	15	15	16	16	17	17	18	18	11	14	10	13	13	14	14	15	15	16	6	7	6	7	8	8	10	11	12	13	0
DF.s	50	16	10	21	2	5	42	11	43	37	31	11	1	36	5	37	13	59	31	6	42	5	33	16	46	58	3	7	7	13
AH.a	143	142	141	139	138	137	136	134	133	132	131	129	128	127	126	124	123	122	120	119	118	116	115	114	112	111	109	108	107	105
AH.b	34	21	8	53	39	24	8	51	34	17	2	47	31	14	57	39	20	1	40	20	2	41	24	4	43	21	58	34	10	44
AH.c	21	17	1	49	14	5	23	49	49	12	36	6	12	41	16	29	40	22	51	0	20	1	19	17	5	18	23	42	8	13⌟

⌟ f. 7r S43 f. 9r S45

⌜ f. 7v S43 f. 9v S45

	⌜121	122	123	124	125	126	127	128	129	130	131	132	133	134	135	136	137	138	139	140	141	142	143	144	145	146	147	148	149	150
EJ.d	40	40	40	40	40	40	41	41	41	41	41	41	40	40	40	40	40	40	40	40	40	39	39	39	39	38	38	38	38	38
EJ.m	47	48	50	51	54	57	1	6	11	17	8	0	51	44	36	34	32	32	31	30	14	58	42	27	12	58	43	30	17	4
EJ.s	58	40	46	53	17	34	21	3	22	59	50	5	44	5	48	28	55	6	43	55	7	29	50	34	50	18	51	23	6	53
DF.m	0	1	2	2	3	3	4	5	6	9	8	8	7	7	2	1	0	0	0	17	17	15	15	14	14	14	13	13	12	35
DF.s	42	6	7	24	17	47	42	19	37	9 ×	45	21	39	17	20	33	49	23	48	48	38	34	21	34	32	27	28	17	13	43
AH.a	104	103	101	100	99	97	96	94	93	92	90	89	88	86	85	83	82	81	79	78	76	75	74	73	71	70	69	67	66	64
AH.b	26	8	49	28	6	44	21	56	30	3	45	27	7	46	24	59	34	7	38	8	54	38	22	4	44	23	9	36	10	41
AH.c	37	12	8	13	59	21	1	18	37	32	48	4	20	5	0	56	32	41	40	21	8	58	29	40	49	49	15	18	1	6

	151	152	153	154	155	156	157	158	159	160	161	162	163	164	165	166	167	168	169	170	171	172	173	174	175	176	177	178	179	180
EJ.d	37	36	36	35	35	34	33	33	32	31	30	29	28	26	25	24	23	21	20	18	17	15	13	11	9	7	6	4	2	0
EJ.m	29	52	15	38	0	22	42	3	22	41	28	17	7	56	42	26	7	46	22	58	12	24	34	41	47	51	0	3	3	0
EJ.s	10	49	54	18	49	10	59	33	55	46	3	50	41	9	30	16	53	50	47	51	55	12	12	39	5	50	27	11	4	0
DF.m	36	36	37	37	38	39	39	40	41	41	63	70	70	71	72	74	74	78	80	84	84	110	112	114	115	115	117	120	123	0
DF.s	21	55	36	39	39	9	26	38	38	9	43	13	9	32	39	14	14	23	3	3	56	0	33	34	14	24	16	7	4	0
AH.a	63	62	61	60	58	57	56	54	53	52	51	50	49	48	47	47	46	45	44	43	42	42	42	41	41	40	40	39	39	39
AH.b	33	23	12	0	46	30	13	54	32	9	21	31	41	50	58	6	12	17	20	23	58	32	7	41	15	48	22	55	27	0
AH.c	5	42	58	48	26	55	49	8	53	36	8	53	48	50	56	33	29	19	58	21	6	59	50	33	45	57	29	8	58	0⌟

⌟ f. 7v S43 f. 9v S45

Table XVII-C: apparatus criticus

‖ bhaumaśīghraphalaṃ ‖] ‖ atha śīghrakendrāṃśoparibhaumasya śīghraphalam aṃśādi tatodhontaraṃ tatodhaḥ karṇa tatodho gatiphalam ‖ koṭijyācalakendrajeli ‖ yadā śīghrakendraḥ ṣaḍrāśibhyo'dhiko bhavati tadā tasya śīghrakendrasya rāśyādi dvādaśarāśitaḥ pātyaṃśeṣāṃśopariśīghraphalaṃ jñeyaṃ ‖ (f. 6^{r}) S43, bhaumaśīghraphalāny aṃśādīni ‖ (f. 6^{v}), (f. 7^{r}) S43, bhaumasya śīghraphalāni ‖ iti bhaumasya śīghraphalāni ‖ (f. 7^{v}) S43; bhaumaśīghraphalaṃ ‖ (f. 8^{v}) S45, ‖ bhaumaśīghraphalāni ‖ (f. 9^{r}), (f. 9^{v}) S45.

AH.b(3) 36 S43; AH.c(3) 16 S43 DF.s(4) 11 S43; DF.s(5) 11 S43 AH.a(5) 220 S45 AH.c(6) 6 S43 EJ.m(11) 25 S43 AH.c(12) 39 S43 AH.c(14) 34 S43 EJ.d(15) 6 S43; AH.b(15) 26 S43 AH.c(17) 17 S43 DF.s(19) 28 S43; AH.c(19) 10 S43 EJ.s(20) 15 S43 AH.c(22) 6 S43 Df.s(23) 49 S43 DF.s(26) 21 S43 EJ.m(32) 36 S43 EJ.s(33) 42 S43 AH.c(38) 28 S43 EJ.m(39) 19 S43 DF.s(40) 56 S43 EJ.s(41) 55 S43 EJ.s(44) 34 S43; AH.s(44) 185 S43; AH.c(44) 17 S43 AH.b(46) 27 S43; AH.c(46) 59 S43 AH.b(47) 47 S43; AH.c(47) 0 S43 AH.b(48) 6 S43; AH.c(48) 5 S43 DF.s(49) 21 S43; AH.b(49) 25 S43; AH.c(49) 6 S43 AH.a(50) 180 S43; AH.b(50) 43 S43; AH.c(50) 23 S43 AH.a(51) 179 S43; AH.b(51) 1 S43; AH.c(51) 5 S43 AH.b(52) 18 S43; AH.c(52) 23 S43 AH.a(53) 177 S43; AH.b(53) 35 S43; AH.c(53) 21 S43 AH.a(54) 176 S43; AH.b(54) 52 S43; AH.c(54) 21 S43 AH.a(55) 176 S43; AH.b(55) 10 S43; AH.c(55) 15 S43 AH.a(56) 175 S43; AH.b(56) 28 S43; AH.c(56) 25 S43 AH.a(57) 174 S43; AH.b(57) 45 S43; AH.c(57) 30 S43 AH.a(58) 174 S43; AH.b(58) 3 S43; AH.c(58) 35 S43 DF.s(59) 18 S43; AH.a(59) 173 S43; AH.b(59) 21 S43; AH.c(59) 45 S43 AH.a(60) 172 S43; AH.b(60) 45 S43; AH.c(60) 10 S43 EJ.s(61) 13 S43; AH.a(61) 172 S43 AH.a(62) 171 S43 AH.b(63) 27 S43; AH.c(63) 34 S43 DF.s(64) 17 S43; AH.b(64) 26 S43 DF.s(65) 39 S43 AH.c(77) 48 S43 EJ.m(82) 14 S43 AH.c(83) 43 S43 DF.s(86) 20 S43; AH.c(86) 52 S43 DF.m(87) 22 S43; AH.c(87) 43 S43 DF.m(88) 23 S43 DF.m(90) 14 S43; AH.c(90) 45 S43 AH.c(92) 27 S43 DF.m(94) 17 S43 AH.c(97) 13 S43 AH.a(99) 132 S43 AH.c(101) 34 S43 AH.a(105) 125 S43 DF.s(106) 27 S43; AH.c(106) 19 S43 AH.c(107) 41 S43 AH.c(108) 12 S43 AH.c(109) 55 S43 EJ.s(118) 31 S43 AH.b(119) 1 S43 AH.b(121) 36 S43 EJ.m(122) 58 S43 EJ.m(123) 49 S43 AH.c(126) 12 S43 DF.m(129) 6 dha S43 DF.m(130) 9 × S43 AH.b(135) 14 S43 AH.c(137) 22 S43 EJ.s(139) 3 S43; DF.s(139) 38 S43 DF.m(140) 16 S43 DF.m(141) 15 S43 EJ.s(142) 59 S43 EJ.s(143) 55 S43 DF.s(144) 44 S43 AH.b(147) 1 S43; AH.c(147) 35 S43 AH.b(148) 38 S43; AH.c(148) 8 S43 AH.c(150) 16 S43 AH.b(152) 33 S43 DF.s(153) 29 S43 DF.s(154) 29 S43 AH.b(155) 48 S43 DF.s(156) 11 S43 DF.s(157) 22 S43; AH.c(157) 44 S43 EJ.s(158) 37 S43 DF.s(159) 9 S43 EJ.m(160) 73 S43; EJ.s(160) 43 S43 EJ.m(161) 70 S43; EJ.s(161) 13 S43; AH.c(161) 53 S43 EJ.s(162) 9 S43 EJ.m(163) 71 S43; EJ.s(163) 32 S43, AH.b(163) 1 S43; AH.c(163) 49 S43 DF.m(164) 73 S43; DF.s(164) 39 S43 DF.m(165) 76 S43; DF.s(165) 14 S43; AH.b(165) 48 S43 DF.m(166) 78 S43; DF.s(166) 23 S43 DF.m(167) 71 S43; DF.s(167) 3 S43 DF.m(168) 74 S43; DF.s(168) 3 S43 DF.m(169) 83 S43; DF.s(169) 56 S43 DF.m(170) 106 S43; DF.s(170) 56 S43 EJ.s(171) 15 S43; DF.m(171) 108 S43; DF.s(171) 3 S43 DF.s(175) 34 S43 EJ.s(176) 51 S43; AH.b(176) 40 S43 EJ.s(177) 7 S43 AH.c(178) 57 S43 AH.c(179) 57 S43.

"‖ śīghrakarṇasya sphuṭīkaraṇaṃ śīghrakesyaṃ śamānena koṣṭakasvakarṇogrāhyāḥ tataḥ koṣṭadvayāntaraṃ śīghrakendrasya saṃguṇya ṛṇaṃkalāvikalābhyāṃ kāryaṃ sphuṭakaṃ ‖" paratext in the left margin of f. 8^{v} S45; "‖ śīghrakendre ṣaṭrāśibhyo nyūne atīśīghrakendraṃśopariśīghraphalaṃ koṣṭakasthaṃ antaraṃ kalādinā saṅguṇya yute sphuṭaṃ śīghraphalaṃ ṣaṭrāsibhyo'dhike dvādeśaśuddhāṃśopariphalaṃ adhaḥ śīghrakarṇaḥ śīghraphalaṃ kendravaśād dhanarṇa meṣādikendredhanaṃ tulādikendre ṛṇaṃ ‖" paratext in the right margin of f. 8^{v} S45 • Table XVII-C *om.* SMB.

Table XVIII-A

⌜f. 17ᵛ B	⌜atha budhaśīghraphalamḥ ‖																															
	1	2	3	4	5	6	7	8	9	10	11	12	13	14	15	16	17	18	19	20	21	22	23	24	25	26	27	28	29	30		
	0	0	0	1	1	1	1	2	2	2	2	3	3	3	3	4	4	4	5	5	5	5	6	6	6	6	7	7	7	7		EJ.d
	16	32	48	4	20	36	52	9	25	41	57	12	28	45	59	12	27	46	2	17	32	48	0	19	34	50	5	21	36	52		EJ.m
	7	14	21	28	36	46	57	7	19	36	22	34	14	14	22	52	28	24	19	58	54	20	49	10	33	5	34	12	45	27		EJ.s
	163	163	163	163	163	163	163	163	163	163	163	163	162	162	162	162	162	162	162	162	161	161	161	161	161	160	160	160	160	159		AH.a
	56	53	51	48	44	40	36	12	35	28	19	3	55	47	47	39	31	22	14	6	57	37	22	8	53	53	38	23	8	53⌟	⌟f. 17ᵛ B	AH.b
⌜f. 18ʳ B	⌜31	32	33	34	35	36	37	38	39	40	41	42	43	44	45	46	47	48	49	50	51	52	53	54	55	56	57	58	59	60		
	8	8	8	8	9	9	9	9	10	10	10	10	11	11	11	11	12	12	12	12	13	13	13	13	13	14	14	14	14	15		EJ.d
	6	21	35	50	5	19	34	49	4	20	34	49	4	19	33	48	3	18	23	48	2	15	28	41	55	8	23	36	49	3		EJ.m
	51	16	51	29	1	41	37	20	34	17	39	19	3	39	20	20	25	20	38	56	2	52	29	50	17	49	1	9	20	52		EJ.s
	159	159	158	158	158	158	157	157	156	156	156	155	155	155	154	154	153	153	152	152	151	151	151	151	150	150	150	149	149	149		AH.a
	39	19	59	39	19	0	38	18	53	27	2	31	27	2	36	11	45	29	53	27	1	31	31	2	32	2	32	10	32	2		AH.b
⌜f. 18ᵛ B	61	62	63	64	65	66	67	68	69	70	71	72	73	74	⌜75	76	77	78	79	80	81	82	83	84	85	86	87	88	89	90		
	15	15	15	15	16	16	16	16	16	17	17	17	17	17	17	18	18	18	18	18	18	19	19	19	19	19	19	19	20	20		EJ.d
	15	26	39	51	4	15	28	40	53	6	15	24	34	44	45	4	14	25	34	43	53	2	8	18	25	34	42	51	0	9		EJ.m
	29	32	18	33	35	51	9	47	27	21	34	51	44	1	49	8	8	35	38	47	3	59	51	0	44	9	22	26	37	55		EJ.s
	147	147	146	145	144	144	144	143	143	142	142	141	140	139	139	138	138	137	137	136	136	135	134	134	133	132	132	131	130	129		AH.a
	36	1	27	52	18	43	9	34	59	24	48	13	35	58⌟	20	42	0	26	26	47	47	30	50	9	24	46	4	32	40	57	⌟f. 18ʳ B	AH.b
⌜f. 19ʳ B	91	92	93	94	95	96	97	98	99	100	101	102	103	104	105	106	107	108	109	110	111	112	113	114	115	116	117	118	⌜119	120		
	20	20	20	20	20	20	20	20	21	21	21	21	21	21	21	21	21	21	21	21	21	21	21	21	21	21	21	21	21	21		EJ.d
	14	20	26	31	37	43	59	54	1	7	9	12	15	18	21	23	27	30	33	37	35	33	30	29	28	25	24	23	20	19		EJ.m
	55	36	10	59	5	56	2	50	2	47	57	31	19	19	2	56	5	16	30	11	9	29	32	48	21	45	23	0	19	32		EJ.s
	129	128	127	126	126	126	125	124	124	123	122	121	121	120	119	118	118	117	116	115	115	114	113	112	111	110	109	109	108	107		AH.a
	32	32	5	21	0	0	26	53	9	23	38	53	8	22	38	9	24	39	35	18	12	35	49	4	19	38	48	13⌟	17	30	⌟f. 18ᵛ B (continues)	AH.b

Table XVIII-A

(continued)																																
	121	122	123	124	125	126	127	128	129	130	131	132	133	134	135	136	137	138	139	140	141	42	143	144	145	146	147	148	149	150		
	21	21	20	20	20	20	20	20	20	20	20	19	19	19	19	19	18	18	18	18	17	17	17	16	16	16	16	15	15	15		EJ.d
	13	6	59	54	46	39	33	27	19	12	0	48	36	26	12	0	49	34	22	9	51	34	16	57	39	21	2	43	24	40		EJ.m
	0	36	39	50	59	35	22	3	31	46	17	40	53	53	55	6	28	30	27	47	57	17	22	50	50	11	17	6	25	27		EJ.s
	107	106	105	105	104	103	102	101	100	100	99	98	97	97	96	95	95	94	93	93	92	91	91	90	90	89	88	87	87	86		AH.a
	30	43	56	42	42	42	59	31	47	2	14	32	46	6	25	43	2	20	18	5	16	58	30	46	11	36	26	49	12	36		AH.b
⌜ f. 19v B	151	152	153	154	155	156	157	158	159	⌜160	161	62	163	164	165	166	167	168	169	170	171	172	173	174	175	176	177	178	179	180		
	14	14	13	13	12	12	12	11	11	10	10	9	9	8	8	7	7	6	6	5	5	4	4	3	2	2	1	1	0	0		EJ.d
	39	15	50	24	58	33	7	40	14	46	16	45	14	45	16	45	15	44	13	42	8	34	0	26	52	18	43	9	34	0		EJ.m
	45	0	6	41	58	16	17	36	2	3	11	8	18	8	8	36	9	39	36	17	34	38	31	27	23	7	45	15	45	0		EJ.s
	85	85	84	84	83	83	83	82	82	81	81	80	80	79	79	79	79	78	78	78	78	77	77	77	77	76	76	76	76	76	⌟a f. 19r B	AH.a
	59	22	45	17	49	20	20	52	23⌟a	84	25	56	26	57	40	24	7	50	33	12	0	43	26	9	2	53	49	41	34	41⌟b	⌟b f. 19v B	AH.b

Table XVIII-B

⌜ f. 9ᵛ Kh	⌜1	2	3	4	5	6	7	8	9	10	11	12	13	14	15	16	17	18	19	20	21	22	23	24	25	26	27	28	29	30		
	0	0	0	1	1	1	1	2	2	2	2	3	3	3	3	4	4	4	5	5	5	5	6	6	6	6	7	7	7	7		EJ.d
	16	32	49	4	20	36	52	9	24	41	57	12	28	43	59	15	30	46	2	17	33	46	30	19		50	5	21	36	51		EJ.m
	6	13	21	28	37	46	57	7	54	30	10	34	9	37	23	3	44	26	10	56	11	27	46	10	4	4	36	10	47	46		EJ.s
	162	163	63	63	63	63	63	63	63	63	163	163	62	62	62	62	62	62	62	62	162	161	161	160	160	160	60	160	159	59		AH.a
	57	53	50	43	41	37	34	31	28	26	10	0	55	52	47	39	31	23	15	7	52	23	13	6	51	31	23	10	53	39		AH.b
⌜ f. 10ʳ Kh	31	32	33	34	35	36	37	38	39	40	41	42	43	44	45	46	47	48	49	50	51	52	53	54	55	56	⌜57	58	59	60		
	8	8	8	8	9	9	9	9	10	10	10	10	11	11	11	11	12	12	12	12	13	13	13	13	13	14	14	14	14	15		EJ.d
	6	11	35	50	5	20	34	49	4	20	30	49	3	18	33	46	3	18	33	48	1	14	28	41	55	8	22	35	49	3		EJ.m
	51	19	50	25	4	46	33	24	30	37	17	17	53	37	35	13	18	33	33	47	59	59	15	34	33	33	11	54	46	42		EJ.s
	59	158	58	58	57	57	57	56	56	56	55	55	55	55	55	54	54	54	52	52	51	50	150	15	49	49	49	48	47	47		AH.a
	14	59	39	53	39	29	22	59	39	23	54	39	32	24	27	42	19	17	48	20	52	32	22	23	33	3⌟	33	14	32	2	⌟ f. 9ᵛ Kh	AH.b
	61	62	63	64	65	66	67	68	69	70	71	72	73	74	75	76	77	78	79	80	81	82	83	84	85	86	87	88	89	90		
	15	15	15	15	16	16	16	16	16	17	17	17	17	17	17	18	18	18	18	18	18	19	19	19	19	19	19	19	20	20		EJ.d
	15	27	39	51	3	10	28	40	53	6	15	25	34	44	54	4	14	24	34	35	52	0	9	17	25	34	42	51	0	9		EJ.m
	27	19	16	20	20	49	14	42	16	11	35	4	39	22	10	7	9	28	38	4	56	56	3	15	37	6	42	25	19	4		EJ.s
	46	45	44	44	44	43	42	42	41	41	40	39	39	38	38	37	36	36	35	34	34	32	33	32	31	30	29	29	28	27		AH.a
	27	6	59	44	18	42	59	42	49	13	36	58	20	42	4	4	48	9	30	51	11	20	47	5	23	40	58	15	33	5		AH.b
	91	92	93	94	95	96	97	98	99	100	101	102	103	104	105	106	107	108	109	110	111	112	113	114	115	116	117	118	119	120		
	20	20	20	20	20	21	21	21	21	21	21	21	21	21	21	21	21	21	21	21	21	21	21	21	21	21	21	21	21	21		EJ.d
	15	20	25	31	31	42	48	24	1	5	10	15	18	20	23	37	30	33	37	36	35	33	23	29	27	25	14	22	21	19		EJ.m
	1	24	59	28	6	56	52	58	4	54	35	16	15	35	58	5	13	33	40	58	0	18	18	29	51	58	18	40	5	32		EJ.s
	27	126	125	124	124	23	22	21	120	119	18	18	117	117	115	15	14	13	12	12	12	11	10	9	9	8	107	106	105	105		AH.a
	5	22	37	53	43	24	38	53	20	38	19	0	40	24	40	8	22	36	8	49	5	24	24	49	3	17	34	44	44	43		AH.b

(continues)

Table XVIII-B

(continued)	121	122	123	124	125	126	127	128	129	130	131	132	133	134	135	136	137	138	139	140	141	142	143	144	145	146	147	148	149	150		
	21	21	20	20	20	20	20	20	20	20	20	19	19	19	19	19	18	18	18	18	17	17	17	16	16	16	16	15	15	15		EJ.d
	12	6	59	53	46	40	32	26	19	12	0	6	36	24	12	0	47	35	22	9	12	34	16	58	39	20	1	42	23	3		EJ.m
	59	23	23	13	44	14	13	14	18	39	26	27	21	25	18	2	38	27	37	41	1	13	10	1	39	58	33	33	15	48		EJ.s
	104	103	102	102	101	100	100	99	98	98	97	96	95	94	93	93	92	92	91	90	90	89	89	88	87	87	86	86	85	84		AH.a
	26	43	43	16	32	48	24	48	32	10	60	26	2	20	38	38	56	13	30	46	12	36	1	25	49	13	30	0	22	45		AH.b
⌜f. 10v Kh	151	152	153	154	155	⌜156	157	158	159	160	161	162	163	164	165	166	167	168	169	170	171	172	173	174	175	176	177	178	179	180		
	14	14	13	13	13	12	12	11	11	10	10	9	9	8	8	7	7	6	6	5	5	4	4	3	2	2	1	1	0	0		EJ.d
	39	15	49	24	19	33	7	40	13	45	16	45	16	46	16	45	15	44	13	42	8	34	0	26	52	18	43	9	34	0		EJ.m
	50	31	57	40	30	13	9	39	55	55	7	54	5	8	0	39	6	23	23	10	21	34	36	29	22	7	44	15	40	0		EJ.s
	84	83	92	92	921	81	81	80	80	80	79	79	79	78	78	78	78	77	77	77	77	76	76	76	76	76	76	76	76	76	⌟a f. 10r Kh	AH.a
	18	49	21	52	33⌟a	54	25	51	27	7	40	24	2	51	34	17	17	43	26	9	2	55	48	41	34	27	21	14	7	7⌟b	⌟b f. 10v Kh	AH.b

Table XVIII-C

⌜ f. 8ʳ S43
f. 10ᵛ S45

⌜‖ budhaśīghraphalāni ‖

1	2	3	4	5	6	7	8	9	10	11	12	13	14	15	16	17	18	19	20	21	22	23	24	25	26	27	28	29	30	
0	0	0	1	1	1	1	2	2	2	2	2	3	3	3	4	4	4	5	5	5	5	6	6	6	6	7	7	7	7	EJ.d
16	32	48	4	20	36	52	9	24	41	57	12	28	43	59	15	30	46	2	17	33	48	3	19	34	50	5	21	36	52	EJ.m
6	13	21	28	37	46	57	7	54	30	1	34	9	37	23	3	44	26	10	56	11	27	47	10	35	4	36	10	47	27	EJ.s
16	16	16	16	16	16	16	15	16	15	15	15	15	15	15	15	15	15	15	15	15	15	15	15	15	15	15	15	15	15	DF.m
7	7	7	9	9	11	10	47	28	31	33	25	25	36	41	41	42	44	46	15	18	20	23	25	29	32	34	37	40	24	DF.s
163	163	163	163	163	163	163	163	163	163	163	163	163	162	162	162	162	162	162	162	161	161	161	161	160	160	160	160	159	159	AH.a
56	53	50	47	43	40	37	34	31	27	19	11	3	55	47	39	31	23	14	6	52	37	22	8	53	38	23	8	53	39	AH.b
46	33	20	7	54	40	28	14	1	48	45	42	39	23	18	12	6	0	0	48	0	20	40	0	6	20	35	49	51	0	AH.c

31	32	33	34	35	36	37	38	39	40	41	42	43	44	45	46	47	48	49	50	51	52	53	54	55	56	57	58	59	60	
8	8	8	8	9	9	9	9	10	10	10	10	11	11	11	11	12	12	12	12	13	13	13	13	13	14	14	14	14	15	EJ.d
6	21	35	50	5	19	34	49	4	20	34	49	3	18	33	48	3	18	33	48	1	14	28	41	55	8	22	35	49	3	EJ.m
51	19	50	25	4	46	33	24	30	17	43	17	53	37	25	19	18	23	33	47	51	59	14	33	0	33	11	54	46	42	EJ.s
14	14	14	14	14	14	14	14	14	14	14	14	14	14	14	14	15	15	15	13	13	13	13	13	13	13	13	13	13	13	DF.m
28	31	35	39	42	47	51	16	47	26	36	16	44	48	54	59	5	10	14	4	8	15	19	27	33	38	43	50	56	45	DF.s
159	158	158	158	157	157	157	156	156	156	155	155	155	154	154	153	153	152	152	152	151	151	150	150	149	149	148	148	147	147	AH.a
19	59	39	19	59	39	19	58	38	18	53	27	2	36	11	45	19	53	27	1	32	2	32	2	32	2	32	2	31	1	AH.b
12	12	17	21	14	11	8	55	42	38	4	21	18	34	1	11	28	31	35	45	3	31	37	54	48	40	34	27	56	37⌟	AH.c

⌟ f. 8ʳ S43
f. 10ᵛ S45

⌜ f. 8ᵛ S43
f. 11ʳ S43

⌜61	62	63	64	65	66	67	68	69	70	71	72	73	74	75	76	77	78	79	80	81	82	83	84	85	86	87	88	89	90	
15	15	15	15	16	16	16	16	16	17	17	17	17	17	17	18	18	18	18	18	18	19	19	19	19	19	19	19	20	20	EJ.d
15	27	59	51	3	15	28	40	53	6	15	25	34	44	54	4	14	24	34	45	52	0	9	17	25	34	42	51	0	9	EJ.m
27	19	16	20	30	49	14	45	25	11	35	4	39	22	10	8	9	20	38	4	56	56	3	15	37	6	42	25	19	48	EJ.s
11	11	12	12	12	12	12	12	12	9	9	9	9	9	9	10	10	10	10	7	7	8	8	8	8	8	8	8	9	5	DF.m
52	57	4	0	19	25	31	40	46	24	29	35	43	48	57	2	11	18	26	52	0	7	12	12	29	36	43	54	29	13	DF.s
146	145	145	144	144	143	142	142	141	141	140	139	139	138	138	137	136	136	135	134	134	133	132	132	131	130	129	129	128	127	AH.a
27	52	18	43	9	34	59	24	48	13	35	58	20	42	4	26	47	8	30	30	9	28	46	4	22	40	57	15	32	48	AH.b
18	55	34	54	18	20	17	12	50	41	57	17	34	32	31	5	46	58	0	53	49	17	36	55	43	21	47	8	2	45	AH.c

(continues)

Table XVIII-C

(continued)

	91	92	93	94	95	96	97	98	99	100	101	102	103	104	105	106	107	108	109	110	111	112	113	114	115	116	117	118	119	120
EJ.d	20	20	20	20	20	20	20	20	21	21	21	21	21	21	21	21	21	21	21	21	21	21	21	21	21	21	21	21	21	21
EJ.m	15	20	25	31	37	42	48	54	1	7	9	12	15	18	20	23	27	30	33	36	35	33	31	29	27	25	24	22	21	19
EJ.s	1	24	59	28	6	56	52	58	4	24	57	35	16	5	53	59	5	13	33	58	0	8	18	29	41	58	18	40	5	32
DF.m	5	5	5	5	5	5	6	6	6	2	2	2	2	2	3	3	3	3	3	1	1	1	1	1	1	1	1	1	1	6
DF.s	23	35	29	38	50	56	6	6	20	33	38	41	49	48	6	6	8	23	25	58_{dha}	$52_{\times}$	50	49	49	43	40	38	35	33	33
AH.a	127	126	125	124	124	123	122	121	121	120	119	118	118	117	116	115	115	114	113	112	112	111	110	109	109	108	107	106	105	105
AH.b	5	1	37	53	8	24	38	53	8	22	38	53	9	24	39	53	8	22	35	49	4	19	34	48	3	17	30	43	56	9
AH.c	26	35	31	13	52	0	52	15	3	3	0	41	19	24	44	47	16	15	47	7	43	42	25	52	16	0	26	34	22	3⌋

⌟ f. 8v S43 f. 11r S45

⌜ f. 9r S43 f. 11v S45

⌜‖ budhaśīghraphalāni ‖

	121	122	123	124	125	126	127	128	129	130	131	132	133	134	135	136	137	138	139	140	141	142	143	144	145	146	147	148	149	150
EJ.d	21	21	20	20	20	20	20	20	20	20	20	19	19	19	19	19	18	18	18	18	17	17	17	16	16	16	16	15	15	15
EJ.m	12	6	59	53	46	39	32	26	19	12	0	48	36	24	12	0	47	35	22	9	52	34	16	58	39	20	1	42	23	3
EJ.s	59	23	46	5	21	44	57	13	27	39	16	27	31	25	8	2	38	5	27	41	1	13	10	1	37	58	33	32	15	48
DF.m	6	6	6	6	6	6	6	6	6	12	11	11	12	12	12	12	12	12	12	17	17	18	18	18	18	19	19	19	19	23
DF.s	37	37	41	41	47	47	44	43	48	23	59	56	7	7	16	24	23	38	46	40	48	3	9	34	39	25	1	17	27	58
AH.a	104	103	102	102	101	100	100	99	98	97	97	96	95	95	94	93	93	92	91	90	90	89	89	88	87	87	86	85	85	84
AH.b	26	42	59	15	31	47	2	17	32	46	6	25	43	2	20	28	55	13	30	46	11	36	1	25	49	49	36	59	22	45
AH.c	6	55	28	54	45	27	54	45	19	31	7	3	45	20	31	17	48	13	0	29	46	16	4	8	5	5	28	43	53	31

	151	152	153	154	155	156	157	158	159	160	161	162	163	164	165	166	167	168	169	170	171	172	173	174	175	176	177	178	179	180
EJ.d	14	14	13	13	12	12	12	11	11	10	10	9	9	8	8	7	7	6	6	5	5	4	4	3	2	2	1	1	0	0
EJ.m	39	15	49	24	59	33	7	40	13	46	16	45	16	48	16	45	15	44	13	42	8	34	0	26	52	18	43	9	34	0
EJ.s	50	3	57	40	3	13	7	39	55	55	7	54	5	8	0	7	7	23	23	10	21	34	36	32	22	7	44	15	40	0
DF.m	24	25	25	25	25	26	26	26	27	30	30	29	29	30	30	30	30	31	31	33	33	33	34	34	34	34	34	34	34	0
DF.s	47	6	17	37	6	6	28	44	0	38	13	54	57	0	21	32	43	0	13	49	47	57	4	10	15	23	29	35	0	0
AH.a	84	83	83	82	82	81	81	80	80	79	79	79	79	78	78	78	78	77	77	77	77	76	76	76	76	76	76	76	76	76
AH.b	17	49	20	52	23	54	25	46	26	57	40	24	7	50	34	17	0	43	26	9	2	55	48	34	27	20	14	7	7	0
AH.c	34	8	51	7	29	26	5	4	44	0	30	0	30	53	10	28	45	4	9	13	27	14	23	40	49	57	6	14	14	0⌋

⌟ f. 9r S43 f. 11v S45

Table XVIII-C: apparatus criticus

‖ buddhaśīghraphalāniḥ ‖] ‖ atha śīghrakendrāṃśoparibudhasya śīghraphalam aṃśādi tatodhontaraṃ kalādi tataḥ karṇas tato gatiphalam ‖ budhasya śīghraphalāni ‖ (f. 8r) S43, budhasya śīghraphalam aṃṣādi sāntaraṃ ‖ budhaśya śīghraphalāni ‖ (f. 8v) S43, budhasya śīghraphalam aṃṣādi sāntaraṃ sakarṇaṃ ‖ iti budhasya śīghraphalāni ‖ (f. 9r) S43; bhaumaśīghraphalāni ‖ (f. 10v), (f. 11v) S45, bhaumaśīghraphalāniḥ ‖ (f. 11r) S45.

AH.c(1) 48 S43 DF.s(2) 8 S43; AH.c(2) 23 S43 DF.m(4) 15 S43; DF.s(4) 59 S43 EJ.s(5) 27 S43; DF.s(5) 19 S43 AH.c(7) 27 S43 DF.s(9) 36 S43 EJ.d(12) 3 S43; DF.s(12) 35 S43 DF.s(13) 28 S43; AH.c(13) 49 S43 DF.s(14) 46 S43; AH.c(14) 33 S43 DF.s(15) 40 S43 AH.c(19) 54 S43 DF.s(21) 16 S43 DF.s(24) 15 S43 AH.c(27) 25 S43 AH.c(29) 1 S43 DF.m(30) 14 S45 EJ.s(36) 26 S43; DF.s(36) 46 S43 DF.m(38) 15 S43; DF.s(38) 6 S43 DF.m(39) 15 S43 AH.b(40) 30 S43; AH.c(40) 32 S43 DF.s(41) 34 S43; AH.c(41) 24 S43 DF.s(42) 36 S43; AH.c(42) 26 S43 EJ.m(44) 28 S43 AH.c(55) 49 S43 EJ.s(59) 44 S43; DF.s(59) 58 S43 DF.m(60) 11 S43 EJ.m(63) 39 S43 DF.s(64) 10 S43 AH.c(67) 57 S43 AH.c(70) 30 S43 AH.c(74) 31 S43 EJ.d(76) 17 S43; EJ.m(76) 44 S43; EJ.s(76) 10 S43 EJ.m(77) 24 S43; EJ.s(77) 20 S43 AH.b(80) 50 S43; AH.c(80) 52 S43 DF.m(81) 8 S43 DF.s(84) 22 S43 AH.c(89) 3 S43 AH.b(92) 21 S43 AH.b(100) 23 S43 EJ.s(102) 32 S43; AH.a(102) 119 S43 EJ.s(104) 15 S43; DF.s(104) 38 S43 DF.m(105) 2 S43 DF.m(106) 2 S43 DF.m(107) 2 S43 DF.m(108) 2 S43 EJ.s(109) 36 S43; DF.m(109) 2 dha S43; DF.s(109) 22 S43 DF.m(110) 1 × S43; DF.s(110) 58 S43 DF.s(112) 49 S43 EJ.s(113) 19 S43 AH.b(117) 31 S43; AH.c(117) 36 S43 DF.s(121) 38 S43 AH.b(122) 45 S43 DF.s(124) 44 S43 DF.s(125) 37 S43 AH.c(126) 20 S43 AH.a(127) 99 S43; AH.b(127) 12 S43 DF.s(128) 46 S43; AH.a(128) 98 S43; AH.b(128) 12 S43; AH.c(128) 54 S43 AH.a(129) 97 S43 DF.m(130) 11 S43; DF.s(130) 43 S43 EJ.s(131) 56 S43; DF.m(131) 12 S43; DF.s(131) 9 S43 EJ.s(132) 47 S43; DF.m(132) 12 S43; DF.s(132) 16 S43 AH.c(134) 31 S43 EJ.s(135) 18 S43 AH.b(136) 38 S43 DF.s(137) 33 S43; AH.a(137) 92 S43 AH.c(141) 40 S43 AH.c(142) 13 S43 DF.s(144) 24 S43 AH.b(146) 13 S43; AH.b(149) 21 S43 DF.m(150) 28 S43 AH.a(154) 83 S43 DF.s(155) 50 S43 AH.b(158) 56 S43 DF.s(159) 10 S43 EJ.s(160) 45 S43; DF.s(160) 39 S43; AH.b(160) 40 S43; AH.c(160) 30 S43 DF.s(162) 49 S43 EJ.m(164) 46 S43; DF.s(164) 8 S43 EJ.s(166) 39 S43; DF.s(166) 33 S43 EJ.s(167) 6 S43 AH.c(169) 6 S43 DF.s(172) 58 S43 DF.m(173) 33 S43; DF.s(173) 57 S43; AH.c(173) 21 S43 EJ.s(174) 39 S43; DF.s(174) 17 S43; AH.b(174) 41 S43; AH.c(174) 31 S43 AH.c(175) 40 S43 AH.b(176) 27 S43; AH.c(176) 40 S43 AH.b(177) 20 S43; AH.c(177) 57 S43 AH.b(178) 14 S43; AH.c(178) 16 S43 DF.s(179) 4 S43.

Table XVIII-C *om.* SMB.

Table XIX-A

⌜a f. 20v B ⌜b f. 21r B	⌜a atha śīghraphalaṃḥ ‖																															
	1	2	3	4	5	6	7	8	9	10	11	12	13	14	15	16	17	18	⌜b19	20	21	22	23	24	25	26	27	28	29	30		
	0	0	0	0	0	0	1	1	1	1	1	1	2	2	2	2	2	2	3	3	3	3	3	3	3	4	4	4	4	4		EJ.d
	9	19	28	38	48	57	7	17	27	36	47	55	6	13	23	32	41	51	2	9	18	27	36	46	55	4	13	22	31	40		EJ.m
	38	36	52	40	18	44	39	31	31	43	48	16	13	52	11	19	51	13	33	33	57	55	1	2	8	12	18	31	38	49		EJ.s
	142	142	142	142	142	142	142	142	142	142	142	142	142	142	142	142	142	142	141	141	141	141	141	141	141	140	140	140	140	140		AH.a
	58	58	58	52	52	48	46	44	42	40	38	35	36	21	16	11	6	2⌟	57	52	43	34	25	27	8	58	52	41	33	24	⌟ f. 20v B	AH.b
	31	32	33	34	35	36	37	38	39	40	41	42	43	44	45	46	47	48	49	50	51	52	53	54	55	56	57	58	59	60		
	4	4	5	5	5	5	5	5	5	6	6	6	6	6	6	6	7	7	7	7	7	7	7	7	7	8	8	8	8	8		EJ.d
	49	56	6	14	22	33	39	48	57	5	13	21	29	36	44	52	0	8	16	26	32	38	45	53	58	5	12	19	26	33		EJ.m
	11	35	0	26	55	23	39	27	1	56	21	35	1	37	49	45	43	46	48	30	5	49	21	3	59	22	51	48	21	51		EJ.s
	140	140	139	139	139	139	139	138	138	138	138	137	137	137	136	136	136	136	135	135	135	135	135	135	134	134	133	133	133	133		AH.a
	12	0	48	36	23	13	0	49	37	25	10	55	48	5	49	49	40	9	54	54	37	19	2	42	12	10	52	35	17	0⌟	⌟ f. 21r B	AH.b
⌜ f. 21v B	⌜61	62	63	64	65	66	67	68	69	70	71	72	73	74	75	76	77	78	79	80	81	82	83	84	85	86	87	88	89	90		
	8	8	8	8	9	9	9	9	9	9	9	9	9	9	9	9	10	10	10	10	10	10	10	10	10	10	10	10	10	10		EJ.d
	39	45	51	56	2	8	14	20	26	32	36	40	45	49	53	57	2	6	10	16	19	21	25	28	31	34	37	40	44	47		EJ.m
	39	22	22	11	53	36	42	40	39	40	41	57	5	21	37	11	1	23	9	11	12	51	6	29	37	33	33	50	28	34		EJ.s
	132	132	132	131	131	131	130	130	130	129	129	128	128	128	127	127	127	126	126	126	125	125	124	124	124	123	123	122	122	122		AH.a
	40	20	0	40	20	0	40	20	0	40	20	57	36	14	53	35	11	49	14	4	41	18	51	32	8	45	51	47	34	12		AH.b
⌜ f. 22r B	91	92	93	94	95	96	97	98	99	100	101	102	⌜103	104	105	106	107	108	109	110	111	112	113	114	115	116	117	118	119	120		
	10	10	10	10	10	10	10	10	10	10	10	10	10	10	10	10	10	10	10	10	10	10	10	10	10	10	10	10	10	10		EJ.d
	48	49	51	53	53	54	55	56	57	59	58	57	57	56	55	55	54	54	53	52	49	46	43	40	36	33	30	27	23	20		EJ.m
	30	51	3	2	52	20	28	42	21	1	26	50	19	32	10	11	31	0	22	44	34	24	22	5	55	35	17	5	50	35		EJ.s
	121	121	120	120	120	119	119	118	118	118	117	117	116	116	116	115	115	115	114	113	113	113	112	112	112	111	111	111	110	110		AH.a
	47	23	58	35	11	37	23	58	31	10	46	22⌟	59	35	12	48	24	0	12	49	26	3	39	39	16	53	29	6	42	18	⌟ f. 21v B (continues)	AH.b

Table XIX-A

(continued) ⌜f. 22v B	121	122	123	124	125	126	127	128	129	130	131	132	133	134	135	136	137	138	139	140	141	42	43	⌜144	145	146	147	148	149	150		
	10	10	10	9	9	9	9	9	9	9	9	9	9	8	8	8	8	8	8	8	7	7	7	7	7	7	7	6	6	6		EJ.d
	15	9	4	58	53	48	42	35	32	26	19	12	4	56	48	40	32	24	17	9	59	50	40	31	20	11	1	51	41	31		EJ.m
	7	38	12	11	35	54	45	25	1	36	1	1	0	11	25	41	54	46	11	16	42	5	21	43	57	12	8	24	26	28		EJ.s
	109	109	109	108	108	108	107	107	107	106	106	105	105	105	105	104	104	104	103	103	103	102	102	102	102	101	101	101	100	100		AH.a
	34	34	14	53	31	10	48	26	4	46	25	3	43	24	5	45	25	5	45	25	8	53	38⌟	21	5	5	32	56	59	43	⌞f. 22r B	AH.b
	151	152	153	154	155	156	157	158	159	160	161	162	163	164	165	166	167	168	169	170	171	172	173	174	175	176	177	178	179	180		
	6	6	5	5	5	5	5	4	4	4	4	4	3	3	3	3	3	2	2	2	2	1	1	1	1	0	0	0	0	0		EJ.d
	19	8	57	45	32	21	9	58	45	33	20	7	54	40	27	14	1	48	24	7	7	53	39	25	10	57	42	29	14	0		EJ.m
	28	16	2	43	54	57	9	34	33	3	8	5	1	54	45	35	23	12	54	29	29	23	14	7	57	23	38	1	12	0		EJ.s
	100	100	100	99	99	99	99	99	98	98	98	98	98	98	98	97	97	97	97	97	97	97	97	97	97	97	97	97	97	97		AH.a
	31	18	6	53	41	28	16	4	51	38	31	24	17	10	3	56	49	42	36	28	25	22	20	17	14	11	8	5	3	0⌟	⌞f. 22v B	AH.b

Table XIX-B

⌜ f. 10^{v} Kh	⌜1	2	3	4	5	6	7	8	9	10	11	12	13	14	15	16	17	18	⌜b19	20	21	22	23	24	25	26	27	28	29	30		
	0	0	0	0	0	0	1	1	1	1	1	1	2	2	2	2	2	2	3	3	3	3	3	3	3	4	4	4	4	4		EJ.d
	9	19	28	38	48	57	7	17	27	36	45	55	4	13	23	32	41	51	0	9	17	27	37	46	55	4	13	22	31	40		EJ.m
	39	18	58	35	18	58	37	21	16	26	59	16	34	52	11	30	51	11	32	44	44	48	14	21	21	18	19	21	22	4		EJ.s
	142	142	142	142	142	142	142	142	142	142	142	142	142	142	142	142	142	142	141	141	141	141	141	141	141	140	140	140	140	140		AH.a
	58	58	51	50	48	46	45	42	41	40	36	35	26	20	14	10	07	00	57	52	43	35	26	17	14	59	50	42	33	24		AH.b

	31	32	33	34	35	36	37	38	39	40	41	42	43	44	45	46	47	48	49	50	51	52	53	54	55	56	57	58	59	60		
	4	4	5	5	5	5	5	5	5	6	6	6	6	6	6	6	7	7	7	7	7	7	7	7	7	8	8	8	8	8		EJ.d
	49	57	5	14	22	31	39	48	56	5	13	21	19	36	44	52	0	8	16	24	31	38	45	52	59	5	12	19	26	33		EJ.m
	11	34	59	18	52	21	50	24	57	32	19	9	1	5	48	45	40	44	45	56	25	23	11	11	53	03	56	45	21	51		EJ.s
	140	140	139	139	139	139	139	138	138	138	177	177	117	136	136	136	136	136	136	13 gha	135	135	135	134	134	134	133	133	133	133		AH.a
	12	01	49	37	25	13	1	49	37	11	55	40	25	10	35	40	34	20	3	50	27	20	31	45	17	52	52	52	55	5		AH.b

⌜ f. 11^{r} Kh	61	62	63	64	65	66	67	68	69	70	71	72	73	74	75	76	77	78	79	80	81	82	83	84	85	86	87	88	⌜89	90		
	8	8	8	8	9	9	9	9	9	9	9	9	9	9	9	9	10	10	10	10	10	10	10	10	10	10	10	10	10	10		EJ.d
	39	45	51	56	2	8	14	19	26	32	36	40	45	49	53	57	2	6	11	15	18	51	25	28	31	34	36	41	44	47		EJ.m
	35	21	7	58	49	44	39	24	36	38	45	54	4	14	39	44	7	38	12	57	52	57	5	10	33	36	49	4	21	39		EJ.s
	13	132	132	131	131	131	131	131	130	130	129	129	129	28	127	127	127	126	126	26	121	125	124	124	123	123	123	122	122	122		AH.a
	40	21	11	10	41	20	10	41	21	10	19	19	56	45	54	32	10	48	26	5	41	18	55	33	8	42	39	58⌟	35	11	⌞ f. 10^{v} Kh	AH.b

	91	92	93	94	95	96	97	98	99	100	101	102	103	104	105	106	17	108	109	110	111	112	112	114	115	116	117	118	119	120		
	10	10	10	10	10	10	10	10	10	10	10	10	10	10	10	10	10	10	10	10	10	10	10	10	10	10	10	10	10	10		EJ.d
	48	49	50	52	53	54	55	56	57	59	58	57	57	56	55	55	54	53	53	52	49	16	43	40	36	33	30	27	23	20		EJ.m
	44	49	55	1	8	18	27	38	50	1	23	44	6	27	48	11	33	55	19	44	33	22	10	3	42	36	11	2	49	30		EJ.s
	122	121	120	120	120	119	119	118	118	118	117	117	116	116	116	115	15	15	14	14	13	113	113	113	12	12	11	11	10	110		AH.a
	47	24	59	36	12	48	23	59	34	10	46	32	59	36	12	48	24	1	36	12	49	26	40	40	17	53	30	40	43	19		AH.b

(continues)

Table XIX-B

(continued)

	121	122	123	124	125	126	127	128	129	130	131	132	133	134	135	136	137	138	139	140	141	142	143	144	145	146	147	148	149	150
EJ.d	10	10	10	9	9	9	9	9	9	9	9	9	9	8	8	8	8	8	8	8	7	7	7	7	7	7	7	6	6	6
EJ.m	15	9	10	58	53	48	42	37	31	26	19	11	3	56	48	40	32	25	17	9	59	50	40	30	20	11	1	51	41	21
EJ.s	0	37	7	41	24	4	45	32	57	36	0	24	43	8	25	51	53	2	9	14	19	4	24	42	56	9	18	24	27	27
AH.a	109	9	9	8	8	8	7	7	7	6	6	106	5	5	5	4	14	4	103	3	3	2	102	102	2	101	101	100	100	11
AH.b	57	37	15	53	32	10	48	26	5	43	23	4	44	28	5	48	25	16	46	26	10	53	37	21	4	48	32	26	59	56

	151	152	153	154	155	156	157	158	159	160	161	162	163	164	165	166	167	168	169	170	171	172	173	174	175	176	177	178	179	180	
EJ.d	6	6	5	5	5	5	5	4	4	4	4	4	3	3	3	3	3	3	3	3	2	2	1	1	1	0	0	0	0	0	
EJ.m	19	8	56	44	32	21	9	57	45	43	20	7	53	40	27	14	1	48	34	21	7	56	39	35	10	56	39	28	12	0	
EJ.s	39	9	36	41	43	1	7	10	18	7	6	3	59	52	44	24	22	8	52	34	29	22	14	6	57	47	48	24	11	0	
AH.a	10	100	100	99	99	99	99	99	98	18	98	98	98	98	18	17	97	97	97	97	97	97	97	97	97	97	97	97	97	97	
AH.b	31	19	6	54	45	29	16	4	51	39	32	25	18	11	4	57	50	42	35	28	26	23	20	17	14	12	9	6	3	0⌟	⌞ f. 11r Kh

Table XIX-C

⌜ f. 9ᵛ S43 f. 12ᵛ S45

⌜‖ guroḥ śīghraphalāni ‖

	1	2	3	4	5	6	7	8	9	10	11	12	13	14	15	16	17	18	19	20	21	22	23	24	25	26	27	28	29	30
EJ.d	0	0	0	0	0	0	1	1	1	1	1	1	2	2	2	2	2	2	3	3	3	3	3	3	3	4	4	4	4	4
EJ.m	9	19	28	38	48	57	7	17	27	36	45	55	4	13	23	32	41	51	0	9	18	27	36	46	55	4	13	22	31	40
EJ.s	39	18	58	35	18	58	37	21	1	26	59	16	34	52	11	30	50	11	32	14	52	54	56	0	4	11	17	25	26	42
DF.m	9	9	9	9	9	9	9	9	9	9	9	9	9	9	9	9	9	9	9	9	9	9	9	9	9	9	9	9	9	8
DF.s	39	40	37	43	40	39	44	40	25	33	17	18	18	19	19	20	21	22	22	58	2	2	4	4	7	6	8	1	16	29
AH.a	142	142	142	142	142	142	142	142	142	142	142	142	142	142	142	142	142	142	141	141	141	141	141	141	141	140	140	140	140	140
AH.b	58	56	54	52	50	48	46	44	42	40	35	31	26	21	16	11	6	2	57	52	43	34	25	17	8	59	50	41	33	24
AH.c	4	8	12	16	20	25	29	43	37	41	52	2	13	23	34	44	55	5	2	10	26	41	56	11	27	29	41	52	4	15

	31	32	33	34	35	36	37	38	39	40	41	42	43	44	45	46	47	48	49	50	51	52	53	54	55	56	57	58	59	60
EJ.d	4	4	5	5	5	5	5	5	5	6	6	6	6	6	6	6	7	7	7	7	7	7	7	7	7	8	8	8	8	8
EJ.m	49	57	5	14	22	31	39	48	56	5	13	21	29	36	44	52	0	8	16	24	31	38	45	52	58	5	12	19	26	33
EJ.s	11	34	59	18	52	21	20	24	57	32	19	9	1	54	48	45	43	44	45	50	35	23	11	3	56	51	48	47	48	51
DF.m	8	8	8	8	8	8	8	8	8	7	7	7	7	7	7	7	8	8	8	6	6	6	6	6	6	7	6	7	7	5
DF.s	23	25	29	34	29	29	34	33	35	47	50	52	53	54	57	58	1	1	5	45	48	48	52	53	45	7	59	1	3	44
AH.a	140	140	139	139	139	139	139	138	138	138	138	137	137	137	137	136	136	136	136	135	135	135	135	134	134	134	133	133	133	133
AH.b	12	0	48	36	25	13	1	49	37	25	10	15	40	25	10	55	40	24	9	54	37	19	2	45	27	10	52	35	17	0
AH.c	30	33	43	54	4	14	24	17	22	28	34	26	26	26	26	10	4	57	51	39	14	59	49	9	46	24	44	14	43	0⌟

⌟ f. 9ᵛ S43 f. 12ᵛ S45

⌜ f. 10ʳ S43 f. 13ʳ S45

	⌜61	62	63	64	65	66	67	68	69	70	71	72	73	74	75	76	77	78	79	80	81	82	83	84	85	86	87	88	89	90
EJ.d	8	8	8	8	9	9	9	9	9	9	9	9	9	9	9	9	10	10	10	10	10	10	10	10	10	10	10	10	10	10
EJ.m	39	45	51	56	2	8	14	19	26	32	36	40	45	49	53	47	2	6	11	15	18	21	25	28	31	34	37	41	44	47
EJ.s	35	20	7	58	49	44	39	24	36	38	45	54	4	14	29	44	7	38	12	47	52	57	5	10	23	36	49	4	21	39
DF.m	5	5	5	5	5	5	5	7	6	4	4	4	4	4	4	4	4	4	4	3	3	3	3	3	3	3	3	3	3	1
DF.s	45	47	51	51	55	55	45	12	2	7	9	10	10	15	15	23	31	34	45	1	5	8	5	13	13	14	15	17	18	5
AH.a	132	132	132	131	131	131	130	130	130	129	129	128	128	128	128	127	127	126	126	126	125	125	124	124	124	123	123	122	122	122
AH.b	40	20	0	40	20	0	40	20	0	40	19	57	36	14	53	31	10	48	26	4	41	18	55	32	8	45	21	58	34	11
AH.c	17	34	51	46	54	49	48	47	46	23	9	40	16	51	12	38	4	19	37	43	31	31	14	38	52	20	58	20	46	13

(continues)

Table XIX-C

(continued)

	91	92	93	94	95	96	97	98	99	100	101	102	103	104	105	106	107	108	109	110	111	112	113	114	115	116	117	118	119	120
EJ.d	10	10	10	10	10	10	10	10	10	10	10	10	10	10	10	10	10	10	10	10	10	10	10	10	10	10	10	10	10	10
EJ.m	48	49	50	52	53	54	55	56	57	59	58	57	57	56	55	55	54	53	53	52	49	46	43	40	36	33	30	27	23	20
EJ.s	44	49	55	1	8	18	27	38	50	1	23	44	6	27	48	11	33	55	19	44	33	21	10	3	46	33	19	2	49	30
DF.m	1	1	1	1	1	1	1	1	1	0	0	0	0	0	0	0	0	0	0	3	3	3	3	3	3	3	3	3	3	5
DF.s	5	6	6	7	10	9	11	12	11	38×	39	38	39	39	37	37	37	36	35	11	11	12	7	7	13	14	17	13	14	29
AH.a	121	121	120	120	120	119	119	118	118	118	117	117	116	116	116	115	115	115	114	114	113	113	113	112	112	111	111	111	110	110
AH.b	47	23	59	35	11	47	23	58	34	10	46	23	59	35	12	48	24	0	36	12	49	26	3	39	16	53	29	6	42	18
AH.c	18	33	33	36	39	21	12	46	26	5	31	8	29	53	38	21	33	46	31	6	22	22	22	25	43	11	40	22	32	55⌟

⌟ f. 10r S43 f. 13r S45

⌜ f. 10v S43 f. 13v S45

⌜‖ guroḥ śīghraphalāni ‖

	121	122	123	124	125	126	127	128	129	130	131	132	133	134	135	136	137	138	139	140	141	142	143	144	145	146	147	148	149	150
EJ.d	10	10	10	9	9	9	9	9	9	9	9	9	9	8	8	8	8	8	8	8	7	7	7	7	7	7	7	6	6	6
EJ.m	15	9	4	58	53	48	42	37	31	26	19	11	3	56	46	40	32	25	17	9	59	50	40	30	20	11	1	51	41	31
EJ.s	1	37	7	41	24	4	45	22	57	33	0	24	48	8	24	41	53	2	9	14	19	4	24	42	56	9	18	24	27	27
DF.m	5	5	5	5	5	5	5	5	5	7	7	7	7	7	7	7	7	7	7	8	10	9	9	9	9	9	9	9	10	11
DF.s	24	30	26	17	20	19	23	25	25	33	36	36	40	44	43	48	51	53	55	55	15	42	42	46	47	51	54	57	1	38
AH.a	109	109	109	108	108	108	107	107	107	106	106	106	105	105	104	104	104	103	103	103	102	102	102	102	102	101	101	101	100	100
AH.b	57	36	14	53	31	10	48	26	4	42	23	3	44	24	5	45	25	5	45	20	9	53	37	21	5	48	32	16	59	43
AH.c	29	9	50	10	39	8	13	30	43	37	16	55	9	37	5	8	25	42	34	40	45	28	23	19	14	45	31	17	45	21

	151	152	153	154	155	156	157	158	159	160	161	162	163	164	165	166	167	168	169	170	171	172	173	174	175	176	177	178	179	180
EJ.d	6	6	5	5	5	5	5	4	4	4	4	4	3	3	3	3	3	2	2	2	2	1	1	1	1	0	0	0	0	0
EJ.m	19	8	56	44	32	21	9	57	45	33	20	7	53	40	27	14	1	48	34	21	7	53	39	25	10	56	42	28	14	0
EJ.s	49	9	26	41	53	1	7	10	10	7	6	3	59	52	44	34	22	8	52	34	29	22	14	6	57	47	38	24	12	0
DF.m	11	11	11	11	11	11	11	12	12	13	13	13	13	13	13	13	13	13	17	17	17	14	14	14	14	14	14	14	14	0
DF.s	40	43	45	48	52	54	57	0	3×	1	3	4	7	8	10	12	14	16	5	5	5	7	6	9	9	4	14	12	12	0
AH.a	100	100	100	99	99	99	99	99	98	98	98	98	98	98	98	97	97	97	97	97	97	97	97	97	97	97	97	97	97	97
AH.b	31	18	6	53	41	28	16	4	51	38	31	24	17	10	3	56	49	42	35	28	25	22	20	17	14	10	8	5	3	0
AH.c	4	46	28	49	24	58	33	8	20	47	49	50	52	54	56	30	35	33	30	28	39	50	1	12	13	35	45	7	7	0⌟

⌟ f. 10v S43 f. 13v S45

Table XIX-C: apparatus criticus

‖ guroḥ śīghraphalāni ‖] ‖ atha śīghrakendrāṃśopariguroḥ śīghraphalam aṃśādi sāntaram adhaḥ karṇādhogatiphalāni ‖ guroḥ śīghraphalāni ‖ (f. 9v); S43, guroḥ śīghraphalam aṃśādīni ‖ guroḥ śīghraphalāni ‖ (f. 10r) S43, śīghrakendrāṃśopariguroḥ śīghraphalam aṃśādi ‖ iti guroḥ śīghrakendrāṃśopari śīghraphalāni ‖ (f. 10v) S43; guroḥ śīghraphalāni ‖ (f. 12v), (f. 13v) S45, guruśīghraphalalāniḥ ‖ (f. 13r) S45.

DF.s(3) 40 S43 EJ.s(4) 38 S43; DF.s(4) 40 S43 EJ.s(6) 59 S43; DF.s(6) 41 S43 EJ.s(7) 40 S43; DF.s(7) 41 S43 DF.s(8) 33 S43 DF.s(9) 43 S43 EJ.s(10) 44 S43; DF.s(10) 15 S43 AH.c(11) 32 S43 AH.c(14) 24 S43 DF.s(17) 28 S43 EJ.s(18) 18 S43; DF.s(18) 14 EJ.s(20) 54 S43; DF.m(18) 8 S43 AH.b(30) 25 S43 DF.s(33) 19 S43 AH.c(34) 56 S43 EJ.s(37) 50 S43; DF.m(40) 8 S43 AH.b(42) 55 S43 AH.b(45) 0 S43 AH.c(48) 7 S43 AH.c(50) 29 S43 AH.c(53) 45 S43 EJ.s(56) 41 S43; AH.c(59) 44 S43 DF.m(60) 8 S43 AH.b(62) 34 S43; AH.c(62) 0 S43 DF.m(67) 4 S43; AH.b(67) 42 S43 DF.s(70) 16 S43 EJ.s(71) 54 S43; DF.s(71) 0 S43 AH.a(75) 127 S43 EJ.m(76) 57 S43 DF.s(77) 41 S43; AH.c(77) 40 S43 DF.s(78) 38 S43 EJ.s(79) 16 S43; DF.s(79) 41 S43; AH.c(79) 27 S43 EJ.s(80) 57 S43 EJ.s(81) 58 S43; AH.c(81) 20 S43 DF.s(86) 13 S43; AH.b(86) 55 S43 AH.a(87) 128 S43 AH.b(95) 16 S43 AH.c(98) 48 S43 DF.m(99) 1 dha S43; DF.s(99) 12 S43 EJ.s(100) 2 S43; DF.m(100) 0 × S43 AH.c(105) 18 S43 DF.s(106) 38 S43 AH.c(108) 43 S43 EJ.s(112) 22 S43 DF.s(114) 17 S43; AH.b(114) 36 S43; AH.c(114) 43 S43 AH.c(115) 49 S43 AH.c(117) 47 S43 DF.s(119) 19 S43 AH.c(128) 10 S43 DF.s(129) 24 S43; AH.c(129) 46 S43 AH.c(130) 10 S43 EJ.m(135) 48 S43; AH.a(135) 105 S43 AH.c(136) 5 S43 AH.a(138) 104 S43 DF.m(140) 9 S43; AH.b(140) 25 S43 AH.a(141) 103 S43 EJ.m(142) 49 S43; DF.s(142) 40 S43 DF.s(143) 40 S43 EJ.s(144) 44 S43; DF.s(144) 40 S43 AH.b(147) 12 S43 AH.b(148) 6 S43 DF.s(149) 0 S43 AH.b(150) 53 S43 DF.s(159) 3 S43 AH.c(160) 37 S43 AH.c(166) 37 S43 DF.m(169) 13 S43; DF.s(169) 18 S43 DF.m(170) 14 S43 DF.m(171) 14 S43; DF.s(171) 7 S43; AH.c(171) 38 S43 DF.s(172) 8 S43 DF.s(175) 10 S43; AH.c(175) 33 S43 DF.s(176) 9 S43; AH.b(176) 11 S43; AH.c(176) 34 S43 AH.c(178) 56 S43.

Table XIX-C *om.* SMB.

Table XX-A

	⌜a śukraśīghraphalaṃ ǁ																															
⌜a f. 23^{v} B ⌜b f. 24^{r} B	1	2	3	4	5	⌜b6	7	8	9	10	11	12	13	14	15	16	17	18	19	20	21	22	23	24	25	26	27	28	29	30		
	0	0	1	1	2	2	2	3	3	4	4	5	5	5	6	6	7	7	7	8	8	9	9	9	10	10	11	11	12	12		EJ.d
	25	50	14	41	6	31	57	22	27	13	37	2	26	41	15	40	5	29	54	19	43	7	32	56	22	45	13	37	6	33		EJ.m
	13	28	15	2	21	42	5	29	54	52	37	2	34	7	37	20	4	40	36	57	37	51	12	19	10	45	53	0	7	24		EJ.s
	206	206	206	206	206	206	206	206	205	205	205	205	205	205	204	204	204	204	204	204	203	203	203	202	202	201	201	200	200	200		AH.a
	54	50	44	39	32⌟	28	24	14	9	56	44	30	18	5	53	40	57	20	25	2	39	16	52	28	6	43	20	56	23	9	⌟ f. 23^{v} B	AH.b
⌜ f. 24^{v} B	31	32	33	34	35	36	37	38	39	40	41	42	43	44	45	46	47	48	49	50	51	52	⌜53	54	55	56	57	58	59	60		
	12	13	13	14	14	15	15	15	16	16	17	17	17	18	18	19	19	19	20	20	21	21	21	22	22	23	23	24	24	24		EJ.d
	56	21	45	10	35	0	25	50	15	40	5	28	53	17	41	6	31	54	22	46	12	36	59	23	46	12	37	1	27	52		EJ.m
	39	6	52	14	10	3	10	24	37	37	31	54	0	18	40	24	7	14	32	53	35	8	38	34	21	22	31	54	4	38		EJ.s
	199	199	199	198	197	197	196	196	195	198	194	194	192	192	192	192	190	190	189	188	188	187	186	185	184	184	183	182	181	180		AH.a
	38	7	5	4	32	0	28	56	24	8	52	12	32	51	10	30	39	8	28	45	3	16⌟	28	56	53	5	16	16	39	50	⌟ f. 24^{r} B	AH.b
	61	62	63	64	65	66	67	68	69	70	71	72	73	74	75	76	77	78	79	80	81	82	83	84	85	86	87	88	89	90		
	25	25	25	26	26	27	27	27	28	28	29	29	29	30	30	30	31	31	32	32	32	33	33	33	34	34	34	35	35	36		EJ.d
	24	26	29	22	45	8	32	56	20	45	4	24	43	3	26	49	12	37	0	26	46	6	27	49	11	34	56	20	44	8		EJ.m
	24	43	6	2	3	40	14	19	32	1	13	20	19	20	34	49	56	19	59	54	27	44	45	44	13	35	29	4	35	29		EJ.s
	180	179	178	177	176	175	174	173	172	171	170	169	168	167	166	164	163	162	161	160	158	157	156	155	154	153	151	150	149	149		AH.a
	1	6	10	15	18	22	25	38	31	33	35	33	34	34	32	3	18	14	9	4	55	46	36	26	15	53	22	39	21	13		AH.b
⌜ f. 25^{r} B	91	92	93	94	95	96	⌜97	98	99	100	101	102	103	104	105	106	107	108	109	110	111	112	113	114	115	116	117	118	119	120		
	36	36	37	37	37	37	38	38	38	39	39	39	40	40	40	40	41	41	42	42	42	42	43	43	43	43	43	44	44	44		EJ.d
	24	42	0	18	37	56	16	36	58	19	34	48	5	22	43	1	22	43	6	27	39	20	2	15	29	43	58	15	32	50		EJ.m
	47	18	9	18	24	55	19	49	1	35	50	58	22	50	4	16	42	56	0	54	12	32	35	24	1	28	21	0	10	52		EJ.s
	146	145	144	143	141	140	139	138	136	135	134	132	131	130	129	127	126	124	122	121	120	119	117	116	114	113	112	110	108	107		AH.a
	58	44	28	12	55	37⌟	19	0	40	20	2	44	25	5	42	23	0	2	13	47	26	3	39	14	49	21	53	24	54	22	⌟ f. 24^{v} B (continues)	AH.b

Table XX-A

(continued)
⌜ f. 25v B

	121	122	123	124	125	126	127	128	129	130	131	132	133	134	135	136	137	138	139	140	⌜141	142	143	144	145	146	147	148	149	150	
EJ.d	44	45	45	45	45	45	45	45	46	46	46	46	46	46	46	46	46	46	46	46	46	46	45	45	45	45	44	44	44	44	
EJ.m	54	1	7	14	21	31	40	55	4	18	15	12	10	9	9	11	14	17	22	30	15	0	45	31	19	7	57	48	40	38	
EJ.s	56	16	13	26	53	16	58	58	16	41	10	50	25	54	45	45	34	40	56	54	18	16	19	39	45	51	6	18	57	19	
AH.a	105	104	103	102	100	98	97	95	94	92	91	89	88	87	85	83	83	80	77	75	74	74	73	71	70	68	67	65	63	62	
AH.b	59	34	9	42	15	45	15	45	10	34	9	43	55	45	14	31	6	30	12	50⌟	27	2	2	35	7	37	5	30	53	14	⌟ f. 25r B

śukraśīghraṃ samāpatāḥ ‖

	151	152	153	154	155	156	157	158	159	160	161	162	163	164	165	166	167	168	169	170	171	172	173	174	175	176	177	178	179	180	
EJ.d	43	43	42	42	41	40	40	39	38	38	37	35	34	32	31	29	28	26	25	23	21	19	17	14	12	9	7	5	2	0	
EJ.m	57	18	39	1	21	42	4	29	36	23	3	42	18	4	22	51	25	50	23	27	36	18	1	42	24	55	31	6	32	0	
EJ.s	16	43	22	1	44	48	36	59	52	19	58	21	15	31	23	1	43	20	25	32	46	49	9	39	52	42	13	16	42	0	
AH.a	60	59	58	56	55	54	52	51	49	48	47	46	45	44	43	42	41	40	39	37	37	36	36	36	35	35	34	34	33	33	
AH.b	57	39	20	58	35	10	41	11	38	2	7	11	14	13	16	15	12	8	13	54	26	53	30	2	22	3	15	3	31	31⌟	⌟ f. 25v B

Table XX-B

⌜f. 11^{v} Kh	⌜atha śīghraphalādhaḥ karṇaḥ ‖																														
	1	2	3	4	5	6	7	8	9	10	11	12	13	14	15	16	17	18	19	20	21	22	23	24	25	26	27	28	29	30	
	0	0	1	1	2	2	2	3	3	4	4	5	5	5	6	6	7	7	7	8	8	9	9	9	10	10	11	11	12	12	EJ.d
	25	50	15	41	6	21	5	52	47	13	37	2	26	5	15	41	5	29	54	19	43	7	30	52	22	47	13	39	6	39	EJ.m
	13	26	44	2	20	40	1	24	47	12	36	3	32	6	46	21	4	49	37	29	36	50	10	34	10	49	55	58	6	30	EJ.s
	206	206	206	206	6	6	6	6	6	206	205	5	5	5	205	204	4	4	4	204	203	203	202	2	2	201	201	200	200	200	AH.a
	55	40	45	45	31	30	25	25	15	15	58	44	19	19	5	5	20	28	15	2	29	16	40	7	42	30	20	57	37	10	AH.b
	31	32	33	34	35	36	37	38	39	40	41	42	43	44	45	46	47	48	49	50	51	52	⌜53	54	55	56	57	58	59	60	
	13	13	13	14	14	15	15	15	16	16	17	17	17	18	18	19	19	19	20	21	21	21	21	22	22	23	23	24	24	24	EJ.d
	56	21	45	10	15	0	25	50	15	41	4	28	52	17	41	6	21	56	22	49	12	25	59	32	47	12	36	1	16	52	EJ.m
	36	2	34	15	4	1	7	21	44	16	58	47	58	13	56	18	7	9	31	15	34	54	54	40	49	14	54	49	59	24	EJ.s
	199	99	98	98	97	97	96	95	94	93	93	92	92	92	91	90	90	89	88	88	87	86	85	84	85	83	82	81	80	80	AH.a
	38	7	25	4	32	2	22	27	25	52	32	53	51	11	27	40	8	27	45	4	16	29	41	53	5	7	8	40	52	2	AH.b
	61	62	63	64	65	66	67	68	69	70	71	72	73	74	75	76	77	78	79	80	81	82	83	84	85	86	87	88	89	90	
	25	25	25	26	26	27	27	27	28	28	29	29	29	30	30	30	31	31	32	32	32	33	33	33	34	34	34	35	35	36	EJ.d
	14	36	59	41	44	8	32	26	20	45	4	33	53	3	26	19	12	36	1	25	46	6	37	49	11	33	56	19	43	8	EJ.m
	31	35	6	46	59	22	4	4	24	40	9	62	12	36	19	26	55	45	22	51	4	49	41	8	3	27	24	47	47	18	EJ.s
	79	78	77	76	75	74	74	72	72	70	69	68	67	66	66	64	63	62	162	160	158	157	57	55	54	53	51	50	49	47	AH.a
	6	11	15	19	22	36	29	11	34	45	34	33	31	28	25	29	18	14	9	4	56	46	36	26	16	50	52	40	27	13	AH.b
⌜f. 12^{r} Kh	91	92	93	94	95	96	97	⌜98	99	100	101	12	103	104	105	106	107	108	109	110	111	112	113	114	115	116	117	118	119	120	
	36	36	37	37	37	37	38	38	38	39	39	39	40	40	40	41	41	41	42	42	42	42	42	43	43	43	43	44	44	44	EJ.d
	25	47	0	8	36	56	14	36	57	33	40	48	18	23	42	1	22	43	4	28	39	50	2	14	28	43	58	14	50	50	EJ.m
	5	19	1	12	59	12	59	42	50	20	57	53	55	3	0	36	0	20	46	35	8	19	22	2	29	1	24	41	37	37	EJ.s
	49	45	44	43	42	40	39	38	136	35	34	131	31	30	28	27	26	24	23	121	20	19	17	16	114	13	111	110	8	17	AH.a
	59	44	28	12	55	38	20⌟	0	41	20	3	44	45	50	44	23	1	38	13	48	26	3	36	15	49	22	54	25	54	22	AH.b

⌟f. 11^{v} Kh (continues)

Table XX-B

(continued)

121	122	123	124	125	126	127	128	129	130	131	132	133	134	135	136	137	138	139	140	141	142	143	144	145	146	147	148	149	150		
44	45	45	45	45	45	45	45	46	46	46	46	46	46	46	46	46	46	46	46	46	46	45	45	45	45	44	44	44	44		EJ.d
54	0	6	13	21	30	40	41	4	18	14	12	10	9	9	11	13	17	23	30	15	1	45	32	19	7	57	48	41	38		EJ.m
32	45	51	58	41	48	32	56	15	1	42	12	18	14	36	2	17	27	25	28	4	32	28	11	19	50	5	9	13	51		EJ.s
105	4	3	101	100	98	97	95	94	92	91	89	88	86	85	83	82	80	78	77	75	74	73	71	70	68	67	65	63	62		AH.a
54	22	51	34	50	48	15	43	10	35	10	43	15	45	15	43	8	31	53	13	51	25	3	36	8	37	5	40	53	14		AH.b

151	152	153	154	155	156	157	158	159	160	161	162	163	164	165	166	167	168	169	170	171	172	173	174	175	176	177	178	179	180		
43	43	43	42	42	40	40	39	38	38	37	35	34	32	31	29	28	26	25	23	21	19	17	14	12	9	7	5	2	0		EJ.d
46	27	38	0	20	42	4	39	58	23	3	41	17	51	22	50	25	52	36	47	30	25	0	42	21	55	3	6	25	0		EJ.m
31	31	58	20	14	12	1	32	15	54	32	56	58	52	5	46	6	7	25	14	48	46	15	32	4	34	26	35	41	0		EJ.s
60	59	58	56	55	55	52	51	49	43	47	46	45	44	43	42	41	40	39	38	37	36	35	35	35	35	34	34	33	31		AH.a
58	50	20	52	35	30	42	12	31	21	7	11	14	16	16	15	13	9	3	56	53	49	30	2	2	3	33	3	23	1 ⌟	⌞ f. 12r Kh	AH.b

Table XX-C

⌜f. 11ʳ S43

	⌜atha śīghrakendrāṃśopariśukrasya śīghraphalāny aṃśādini ‖																													
	1	2	3	4	5	6	7	8	9	10	11	12	13	14	15	16	17	18	19	20	21	22	23	24	25	26	27	28	29	30
EJ.d	0	0	1	1	2	2	2	3	3	4	4	5	5	5	6	6	7	7	7	8	8	9	9	9	10	10	11	11	12	12
EJ.m	25	50	15	41	6	31	57	22	47	13	37	2	26	51	15	40	5	29	54	19	43	7	32	56	22	48	13	40	6	32
EJ.s	13	28	44	2	21	41	3	26	50	13	37	3	34	7	42	22	5	49	38	30	38	51	12	37	10	0	55	0	7	24
DF.m	25	25	25	25	25	25	25	25	25	24	24	24	24	24	24	24	24	24	24	24	24	24	24	25	25	25	26	26	26	24
DF.s	15	16	18	19	20	22	23	24	24	24	26	31	33	35	40	43	44	48	52	8	13	21	25	33	50	55	5	7	17	15
AH.a	206	206	206	206	206	206	206	206	206	206	205	205	205	205	205	204	204	204	204	204	203	203	202	202	202	201	201	200	200	200
AH.b	55	50	45	39	34	30	25	19	14	9	56	44	31	18	6	53	40	28	15	2	39	16	52	29	6	43	20	56	33	9
AH.c											47	8	27	47	7	27	47	7	27	20	10	9	56	47	39	16	1	36	13	51
	śukrasya śīghraphalāni ‖																													
	31	32	33	34	35	36	37	38	39	40	41	42	43	44	45	46	47	48	49	50	51	52	53	54	55	56	57	58	59	60
EJ.d	12	13	13	14	14	15	15	15	16	16	17	17	17	18	18	19	19	19	20	20	21	21	21	22	22	23	23	24	24	24
EJ.m	56	21	45	10	35	0	25	50	15	41	5	28	53	17	41	6	31	56	22	49	12	36	59	23	47	12	36	1	27	52
EJ.s	39	2	37	14	5	3	10	24	47	19	1	58	0	18	40	18	7	14	32	22	35	8	45	43	52	14	58	55	4	29
DF.m	24	24	24	24	25	25	25	25	25	23	23	24	24	24	24	24	25	26	26	23	23	23	23	24	24	24	24	25	25	21
DF.s	23	35	37	51	58	7	14	23	32	42	57	2	18	22	38	49	7	18	50	13	33	37	58	9	22	44	57	9	25	56
AH.a	199	199	198	198	197	197	196	195	195	194	194	193	192	192	191	190	190	189	188	188	187	186	185	184	184	183	182	181	180	180
AH.b	38	7	35	4	32	0	28	56	24	52	12	31	51	10	30	49	8	26	45	3	16	28	41	53	5	16	28	39	50	1
AH.c	26	6	31	2	9	39	41	43	46	38	28	28	28	53	0	0	3	41	14	48	28	52	13	19	21	59	9	39	41	39⌟
	⌜bhṛgoḥ śīghraphalāni																													
	61	62	63	64	65	66	67	68	69	70	71	72	73	74	75	76	77	78	79	80	81	82	83	84	85	86	87	88	89	90
EJ.d	25	25	25	26	26	27	27	27	28	28	29	29	29	30	30	30	31	31	32	32	32	33	33	33	34	34	34	35	35	36
EJ.m	14	36	59	22	45	8	32	56	20	45	4	23	43	3	26	49	13	36	1	25	46	6	27	49	11	33	56	20	44	8
EJ.s	25	43	6	3	4	30	12	9	32	13	13	39	18	36	20	36	10	45	13	58	13	45	50	15	14	33	29	4	2	23
DF.m	22	22	22	23	23	23	23	24	24	19	19	19	20	22	23	23	23	24	24	20	20	21	21	21	22	22	23	23	24	16
DF.s	18	13	57	1	26	42	57	23	41	0	26	39	18	44	16	34	35	28	45	15	32	5	25	59	19	56	35	58	21	57
AH.a	179	178	177	176	175	174	173	172	171	170	169	168	167	166	165	164	163	162	161	160	158	157	156	155	154	153	151	150	149	148
AH.b	6	10	15	18	22	25	28	31	33	35	34	32	30	28	25	21	17	13	9	4	55	46	36	26	15	4	52	39	26	13
AH.c	24	51	2	55	30	47	45	25	46	47	24	39	32	1	7	49	34	59	26	28	35	22	25	25	40	24	18	52	52	17

⌟f. 11ʳ S43

⌜f. 11ᵛ S43

(continues)

Table XX-C

(continued)																																
	bhr̥goḥ śīghraphalāni																															
	91	92	93	94	95	96	97	98	99	100	101	102	103	104	105	106	107	108	109	110	111	112	113	114	115	116	117	118	119	120		
	36	36	37	37	37	37	38	38	38	39	39	39	40	40	40	41	41	41	42	42	42	42	43	43	43	43	43	44	44	44		EJ.d
	25	42	0	18	37	56	16	36	58	19	34	48	4	23	42	1	22	43	5	29	39	50	2	15	29	43	58	15	32	50		EJ.m
	20	19	9	19	6	32	20	49	2	39	10	58	59	14	16	41	22	31	34	0	12	32	35	25	1	28	47	0	10	53		EJ.s
	16	17	18	18	19	19	20	21	21	14	14	16	18	19	19	20	21	22	23	10	11	12	12	13	14	15	16	17	18	4		DF.m
	59	50	10	47	26	48	29	13	37	31	48	1	15	2	25	41	9	3	26	12	20	3	50	36	27	19	13	10	43	35		DF.s
	146	145	144	143	141	140	139	138	136	135	134	132	131	130	128	127	126	124	123	121	120	119	117	116	114	113	111	110	108	107		AH.a
	58	44	28	12	55	37	19	0	40	20	2	44	25	5	44	23	0	37	13	47	26	3	39	14	48	21	53	24	54	22		AH.b
	53	1	48	22	21	47	32	36	36	4	40	12	13	29	39	12	56	18	18	57	0	27	28	46	47	57	47	38	10	33⌟	⌞ f. 11^{v} S43	AH.c
⌜ f. 12^{r} S43	⌜śīghrakendrāṃśopariśukrasya śīghraphalam aṃśādiḥ																															
	121	122	123	124	125	126	127	128	129	130	131	132	133	134	135	136	137	138	139	140	141	142	143	144	145	146	147	148	149	150		
	45	45	45	45	45	45	45	45	46	46	46	46	46	46	46	46	46	46	46	46	46	46	45	45	45	45	44	44	44	44		EJ.d
	55	0	7	13	21	31	40	52	4	18	14	12	10	9	5	11	13	17	23	30	15	1	45	32	19	7	57	48	41	35		EJ.m
	28	44	13	27	53	16	57	7	16	10	43	12	25	36	36	2	27	17	25	28	4	32	59	45	45	51	7	25	3	6		EJ.s
	5	6	7	7	9	9	11	12	13	3 ×	2	1	0	4	5 dha	1	3	5	7 ddha	15 ×	13	15	13	13	11	10	8	7	5	37		DF.m
	16	29	14	26	23	41	10	9	ddha 55	28	31	47	49	0 ×	26	52	50	58	13	24	32	33	14	0	54	44	42	22	57	50		DF.s
	105	104	103	101	100	98	97	95	94	92	91	89	88	86	85	83	82	80	78	77	75	74	73	71	70	68	67	65	63	62		AH.a
	59	34	9	42	15	45	15	43	10	34	9	43	15	45	14	41	7	31	52	12	50	27	2	35	7	37	5	30	53	14		AH.b
	4	58	41	53	9	49	25	19	3	50	47	0	11	31	39	47	36	7	47	41	36	21	40	45	36	2	7	16	20	17		AH.c
	iti śukrasya śīghraphalāni ‖																															
	151	152	153	154	155	156	157	158	159	160	161	162	163	164	165	166	167	168	169	170	171	172	173	174	175	176	177	178	179	180		
	43	43	42	42	41	40	40	39	38	38	37	35	34	32	31	29	28	26	25	23	21	19	17	14	12	9	7	5	2	0		EJ.d
	57	18	39	1	21	42	4	29	56	23	3	42	18	51	22	51	25	56	23	47	31	16	1	42	21	55	33	6	35	0		EJ.m
	16	43	23	1	44	21	46	59	22	19	58	22	16	25	23	2	43	39	26	33	47	8	9	39	14	43	28	36	57	0		EJ.s
	38	39	38	39	39	37	34	33	33	79	81	84	86	89	91	85	89	93	95	135	135	134	138	141	145	142	146	150	155	0		DF.m
	33	20	22	17	23	35	47	37	3	21	36	6	51	2	21	19	4	13	53	46	39	59	30	25	31	15	12	39	57	0		DF.s
	60	59	58	56	55	54	52	51	49	48	47	46	45	44	43	42	41	40	39	37	37	36	36	36	35	35	34	34	33	33		AH.a
	57	39	20	58	35	10	41	11	38	2	7	11	14	16	16	15	12	8	3	54	27	58	30	2	31	3	32	3	31	0		AH.b
	44	54	27	44	21	2	46	39	9	27	30	29	22	0	22	21	51	47	0	29	0	42	29	16	30	30	55	5	36	0⌟	⌞ f. 12^{r} S43	AH.c

Table XX-D

⌜ f. 14ᵛ S45

⌜‖ śukraśīghraphalāni ‖

	1	2	3	4	5	6	7	8	9	10	11	12	13	14	15	16	17	18	19	20	21	22	23	24	25	26	27	28	29	30
EJ.d	0	0	1	1	2	2	2	3	3	4	4	5	5	5	6	6	7	7	7	8	8	9	9	9	10	10	11	11	12	12
EJ.m	25	50	15	41	6	31	57	22	47	13	37	2	26	51	15	40	5	29	54	19	43	7	32	56	22	47	13	39	6	32
EJ.s	13	28	44	2	20	40	1	24	47	12	36	3	32	6	42	21	4	49	37	29	36	50	10	34	9	59	55	58	6	21
DF.m	25	25	25	25	25	25	25	25	25	24	24	24	24	24	24	24	24	24	24	24	24	24	25	25	25	25	26	26	26	24
DF.s	15	16	18	18	20	21	23	23	25	24	27	29	34	36	39	43	45	48	52	7	14	20	24	35	50	56	4	8	15	17
AH.a	206	206	206	206	206	206	206	206	206	206	205	205	205	205	205	204	204	204	204	204	203	203	202	202	202	201	201	200	200	200
AH.b	54	49	44	39	34	29	24	19	14	9	56	44	31	18	6	53	40	27	15	2	39	16	52	29	6	43	19	56	33	9
AH.c	57	54	51	49	46	44	41	38	36	33	47	7	47	47	7	16	31	48	4	20	10	8	56	47	39	16	52	35	13	51

	31	32	33	34	35	36	37	38	39	40	41	42	43	44	45	46	47	48	49	50	51	52	53	54	55	56	57	58	59	60
EJ.d	12	13	13	14	14	15	15	15	16	16	17	17	17	18	18	19	19	19	20	20	21	21	21	22	22	23	23	24	24	24
EJ.m	56	21	45	10	35	0	25	50	15	41	4	28	52	17	41	6	31	56	22	49	12	36	59	23	47	12	36	1	26	52
EJ.s	38	2	34	15	4	1	7	21	44	56	57	47	58	13	56	18	7	9	31	15	32	1	44	40	49	14	54	49	59	24
DF.m	24	24	24	24	24	25	25	25	25	23	23	24	24	24	24	24	25	26	26	23	23	23	23	24	24	24	24	25	25	21
DF.s	24	32	41	49	5	6	14	22	32	41	50	11	15	43	22	49	2	22	44	17	29	43	56	9	25	40	55	10	25	57
AH.a	199	199	198	198	197	197	196	195	195	194	194	193	192	192	191	190	190	189	188	188	187	186	185	184	184	183	182	181	180	180
AH.b	38	7	35	3	32	9	28	56	24	52	12	33	51	10	27	49	8	26	45	3	16	28	41	53	5	16	28	39	50	1
AH.c	25	6	31	53	16	38	41	43	45	36	22	0	27	52	23	0	0	41	14	48	28	56	13	19	21	59	25	39	41	39⌟

⌟ f. 14ᵛ S45

⌜ f. 15ʳ S43

⌜‖ śukraśīghraphalāni ‖

	61	62	63	64	65	66	67	68	69	70	71	72	73	74	75	76	77	78	79	80	81	82	83	84	85	86	87	88	89	90
EJ.d	25	25	25	26	26	27	27	27	28	28	29	29	29	30	30	30	31	31	32	32	32	33	33	33	34	34	34	35	35	36
EJ.m	14	36	59	21	44	8	32	56	20	45	4	23	43	3	26	49	12	36	1	25	46	6	27	49	11	33	56	19	53	8
EJ.s	21	35	6	53	59	22	4	4	24	4	9	32	12	36	19	26	55	49	2	51	4	39	41	8	3	27	24	50	47	18
DF.m	22	22	22	23	23	23	24	24	24	19	19	19	20	22	23	23	23	24	24	20	20	21	21	21	22	22	23	23	24	16
DF.s	14	31	47	6	23	42	0	20	40	5	23	40	24	43	7	29	54	23	39	13	35	2	27	55	24	57	26	57	31	47
AH.a	179	178	177	176	175	174	173	172	171	170	169	168	167	166	165	164	164	162	161	160	158	157	156	155	154	153	151	150	149	148
AH.b	6	10	15	18	22	25	28	31	33	35	34	32	30	28	25	21	18	13	9	4	55	46	36	26	15	4	52	39	26	13
AH.c	14	51	2	55	30	46	45	25	46	47	24	39	32	1	7	49	6	59	26	28	35	22	38	25	40	24	20	53	52	17

(continues)

Table XX-D

(continued)

	91	92	93	94	95	96	97	98	99	100	101	102	103	104	105	106	107	108	109	110	111	112	113	114	115	116	117	118	119	120		
	36	36	37	37	37	37	38	38	38	39	39	39	40	40	40	41	41	41	42	42	42	42	43	43	43	43	43	44	44	44		EJ.d
	25	42	0	18	36	56	14	36	57	19	33	48	4	23	42	1	22	43	4	28	39	50	2	15	28	43	58	14	32	50		EJ.m
	5	19	1	12	59	52	59	42	50	38	57	54	55	2	0	36	0	20	46	35	8	19	22	2	39	1	24	41	7	37		EJ.s
	17	17	18	18	19	18	21	21	21	14	14	13	18	18	19	20	21	21	23	10	11	12	12	13	14	15	16	17	18	4		DF.m
	14	42	11	47	53	7	43	8	48	19	57	1	7	58	36	24	20	26	49	33	11	3	40	37	22	23	17	26	30	45		DF.s
	146	145	144	143	141	140	139	138	136	135	134	132	131	130	128	127	126	124	123	121	120	119	117	116	114	113	111	110	108	107		AH.a
	58	44	28	12	55	37	19	0	40	20	2	44	25	5	44	23	0	37	13	47	26	3	39	14	48	21	53	24	54	22		AH.b
	53	0	30	22	21	47	32	23	37	4	40	12	13	29	39	12	54	26	18	57	7	27	27	46	46	56	44	38	6	30⌟	⌟ f. 15^r S45	AH.c
⌜ f. 15^v S45	⌜‖ śukraśīghraphalāni ‖																															
	121	122	123	124	125	126	127	128	129	130	131	132	133	134	135	136	137	138	139	140	141	142	143	144	145	146	147	148	149	150		
	44	45	45	45	45	45	45	45	46	46	46	46	46	46	46	46	46	46	46	46	46	46	45	45	45	45	44	44	44	44		EJ.d
	55	0	5	13	21	30	40	51	4	18	14	12	10	9	9	11	13	17	23	30	15	1	45	32	19	7	57	48	41	34		EJ.m
	22	45	51	58	49	48	42	56	15	11	43	12	18	34	36	2	27	27	15	28	4	32	28	12	16	50	5	9	11	51		EJ.s
	5	6	7	7	8	9	11	12	13	3	2	1	1	1	1	1	4	5	7	15	13	16	13	12	12	10	8	6	6	38		DF.m
	23	36	7	51	51	54	14	19	56$_{dha}$	28$_{\times}$	31	54	40	2	26	25	0	48	13	24	32	4	16	56	11	45	56	58	20	15		DF.s
	105	104	103	101	100	98	97	95	94	92	91	99	88	86	85	83	82	80	78	77	75	74	73	71	70	68	67	65	63	62		AH.a
	59	34	9	42	15	45	15	43	10	34	9	43	15	45	14	41	7	31	52	12	50	25	2	35	7	37	5	30	52	14		AH.b
	12	58	41	52	8	49	25	18	2	50	46	0	10	31	39	47	35	6	47	41	36	21	40	42	36	2	1	16	52	17		AH.c
	151	152	153	154	155	156	157	158	159	160	161	162	163	164	165	166	167	168	169	170	171	172	173	174	175	176	177	178	179	180		
	43	43	42	42	41	40	40	39	38	38	37	35	34	32	31	29	28	26	25	23	21	19	17	14	12	9	7	5	2	0		EJ.d
	56	17	38	0	21	42	4	29	56	22	3	41	17	51	22	50	25	56	23	47	31	15	0	42	21	55	33	6	35	0		EJ.m
	36	54	58	20	24	19	1	25	15	56	32	56	58	25	5	46	6	7	25	14	48	46	55	32	4	34	16	35	41	0		EJ.s
	38	38	38	38	39	38	34	33	33	79	81	56	86	89	91	85	88	92	96	135	135	134	138	141	142	142	146	150	155	0		DF.m
	42	42	56	56	5	18	36	10	15	24	36	51	33	20	19	40	59	42	11	26	2	51	23	28	30	18	41	54	41	0		DF.s
	60	59	58	56	55	54	52	51	49	48	47	46	45	44	43	42	41	40	39	37	37	36	36	36	35	35	34	34	33	33		AH.a
	57	39	20	58	35	10	41	11	38	2	7	11	14	16	16	15	12	8	3	54	27	58	30	2	32	3	32	3	31	0		AH.b
	44	54	26	44	31	2	46	39	9	26	30	29	20	0	21	21	51	46	0	28	0	42	29	16	30	30	54	5	35	0⌟	⌟ f. 15^v S45	AH.c

Table XXI-A

⌜f. 27ʳ B	⌜śaniśīghraphalaṃḥ ‖																															
	1	2	3	4	5	6	7	8	9	10	11	12	13	14	15	16	17	18	19	20	21	22	23	24	25	26	27	28	29	30		
	0	0	0	0	0	0	0	0	0	0	1	1	1	1	1	1	1	1	1	1	2	2	2	2	2	2	2	2	2	2		EJ.d
	5	11	17	23	28	35	41	46	52	58	4	9	15	20	26	32	38	43	50	55	0	5	11	16	22	27	33	39	44	49		EJ.m
	51	43	39	35	21	12	5	58	51	43	21	58	35	58	51	33	9	48	2	5	31	55	22	37	14	42	9	0	6	34		EJ.s
	139	132	132	132	132	132	132	132	132	132	132	132	132	132	132	137	132	132	132	132	132	132	131	131	131	131	131	131	131	131		AH.a
	58	57	56	55	54	53	51	50	49	48	45	42	36	33	37	27	24	21	19	13	8	57	55	22	26	41	40	36	21	25		AH.b
⌜f. 27ᵛ B	31	32	33	34	35	36	37	38	39	40	41	42	43	44	45	46	47	48	⌜49	50	51	52	53	54	55	56	57	58	59	60		
	2	3	3	3	3	3	3	3	3	3	3	3	3	3	4	4	4	4	4	4	4	4	4	4	4	4	4	4	5	5		EJ.d
	58	59	4	9	14	19	24	39	34	39	44	48	52	57	2	6	11	15	20	25	29	33	36	40	44	52	56	56	0	3		EJ.m
	32	31	30	31	30	7	31	30	33	36	8	41	13	48	26	58	34	9	50	28	17	6	58	49	40	26	20	20	15	51		EJ.s
	131	131	131	130	130	130	130	130	130	130	129	129	129	129	129	129	129	129	128	128	128	128	128	128	127	127	127	127	127	127		AH.a
	18	11	4	48	42	35	28	21	14	5	56	47	38	28	28	20	11	2⌟	52	43	33	23	12	2	52	41	31	20	10	0	⌟f. 27ʳ B	AH.b
	61	62	63	64	65	66	67	68	69	70	71	72	73	74	75	76	77	78	79	80	81	82	83	84	85	86	87	88	89	90		
	5	5	5	5	5	5	5	5	5	5	5	5	5	5	5	5	5	5	5	5	5	5	6	6	6	6	6	6	6	6		EJ.d
	7	10	13	16	19	22	26	29	32	35	37	39	41	43	45	48	50	52	54	16	57	58	0	1	2	4	5	6	7	9		EJ.m
	16	23	37	38	50	57	9	16	18	40	44	47	51	57	1	6	12	15	13	31	22	59	23	30	50	5	24	40	9	16		EJ.s
	126	126	126	126	126	125	125	125	125	125	124	124	124	123	123	123	123	123	123	122	122	122	122	122	121	121	121	121	120	120		AH.a
	48	36	35	13	0	49	27	21	14	0	49	37	24	59	59	46	35	22	9	56	43	30	16	3	49	36	22	9	55	42		AH.b
⌜f. 28ʳ B	91	92	93	94	95	96	97	⌜98	99	100	101	102	103	104	105	106	107	108	109	110	111	112	113	114	115	116	117	118	119	120		
	6	6	6	6	6	6	6	6	6	6	6	6	6	6	6	6	6	6	6	6	5	5	5	5	5	5	5	5	5	5		EJ.d
	9	9	9	9	9	9	9	10	10	10	9	8	7	6	5	4	3	3	2	1	58	56	44	52	50	48	45	43	41	38		EJ.m
	22	25	37	42	45	58	51	4	4	24	3	16	29	34	40	49	58	8	17	18	56	46	32	6	1	24	23	24	7	41		EJ.s
	120	120	120	119	119	119	119	118	118	118	118	117	117	117	117	117	116	116	116	116	115	115	115	115	115	114	114	114	114	114		AH.a
	28	25	1	47	33	20	6⌟	52	38	24	12	59	45	32	19	5	52	39	25	12	59	46	32	21	8	44	38	28	16	3	⌟f. 27ᵛ B	AH.b
																															(continues)	

Table XXI-A

(continued) ⌜ f. 28v B																															
	121	22	123	124	125	126	127	128	129	130	131	132	133	134	135	136	137	138	139	140	⌜41	142	143	144	145	146	147	148	149	150	
EJ.d	5	5	5	5	5	5	5	5	5	5	5	4	4	4	4	4	4	4	4	4	4	4	4	3	3	3	3	3	3	3	
EJ.m	35	32	28	25	31	18	15	11	8	4	0	55	51	46	42	27	37	28	23	19	13	8	2	57	52	46	41	35	30	24	
EJ.s	22	2	39	18	43	30	5	42	15	9	31	48	18	30	12	37	32	25	9	10	54	24	59	20	6	37	7	18	7	21	
AH.a	113	113	113	113	112	112	112	112	112	112	111	111	111	111	111	111	110	110	110	110	110	110	109	109	109	109	109	109	109	108	
AH.b	51	38	28	16	5	53	43	29	18	6	55	45	34	24	13	3	52	42	31	21⌟	9	4	55	46	37	29	21	12	4	55	⌞ f. 28r B
	151	152	153	154	155	156	157	158	159	160	161	162	163	164	165	166	167	168	69	70	171	172	173	174	175	176	177	178	179	180	
EJ.d	3	3	3	2	2	2	2	2	2	2	2	2	2	1	1	1	1	1	1	1	1	0	0	0	0	0	0	0	0	0	
EJ.m	18	11	5	59	53	46	40	33	27	21	14	7	0	53	47	40	33	26	19	12	05	58	50	43	36	29	21	14	7	0	
EJ.s	16	58	56	24	03	36	21	58	37	22	23	34	39	54	04	12	21	29	36	44	28	12	56	40	24	07	51	34	17	0	
AH.a	108	108	108	108	108	108	108	108	107	107	107	107	107	107	107	107	107	107	107	107	107	107	107	107	107	107	107	107	107	107	
AH.b	49	43	36	29	23	17	10	4	54	50	47	43	40	36	32	26	25	21	18	14	13	12	10	8	7	6	4	3	2	0⌟	⌞ f. 28v B

"‖ iti śrīkarṇakatūhale mandaśīghraphalaṃ saṃpūrṇaṃ samāptām iti ‖ śubhaṃ astuḥ ‖ saṃvat 1734 varṣe kātī sudī 2 budhavāre pothīlaṣītaṃ caraṃbasu ‖" colophon in the bottom margin of f. 28v B.

Table XXI-B

	⌜a1	2	3	4	5	6	7	8	9	10	11	12	⌜b13	14	15	16	17	18	19	20	21	22	23	24	25	26	27	28	29	30		
⌜a f. 12r Kh ⌜b f. 12v Kh	0	0	0	0	0	0	0	0	0	0	1	1	1	1	1	1	1	1	1	1	2	2	2	2	2	2	2	2	2	2		EJ.d
	5	11	17	23	29	35	41	46	52	58	4	9	15	21	26	32	38	43	49	55	0	5	11	16	22	27	33	38	44	49		EJ.m
	51	43	35	38	20	13	5	58	51	47	21	58	58	23	51	29	8	17	26	5	30	55	21	47	14	41	9	38	4	34		EJ.s
	32	32	32	32	32	32	32	32	32	32	32	32	32	32	32	32	32	32	32	32	32	32	32	31	31	31	31	31	31	31		AH.a
	59	58	56	55	54	52	52	59	44	45	42	39⌟	39	37	34	31	28	25	22	19	13	6	3	58	52	47	42	36	31	26	⌟ f. 12r Kh	AH.b
	31	32	33	34	35	36	37	38	39	40	41	42	43	44	45	46	47	48	49	50	51	52	53	54	55	56	57	58	59	60		
	2	2	3	3	3	3	3	3	3	3	3	3	3	4	4	4	4	4	4	4	4	4	4	4	4	4	4	4	5	5		EJ.d
	54	59	9	14	19	24	29	34	29	44	48	47	57	2	3	6	11	16	20	25	29	33	36	40	44	48	52	56	0	4		EJ.m
	31	29	38	28	29	29	31	33	36	8	41	14	14	43	28	57	34	11	53	57	16	6	56	37	31	32	25	20	14	9		EJ.s
	31	31	31	30	30	30	30	30	30	30	30	29	29	29	29	29	29	29	29	28	28	28	28	28	28	27	27	27	27	27		AH.a
	19	11	4	57	50	43	36	18	15	14	5	56	47	38	33	29	20	11	2	50	44	43	23	12	3	52	42	21	11	1		AH.b
	61	62	63	64	65	66	67	68	69	70	71	72	73	74	75	76	77	78	79	80	81	82	83	84	85	86	87	88	89	90		
	5	5	5	5	5	5	5	5	5	5	5	5	5	5	5	5	5	5	5	5	5	5	6	6	6	6	6	6	6	6		EJ.d
	17	10	13	16	19	22	26	29	32	35	37	39	41	43	45	48	50	52	54	56	57	58	0	1	2	4	5	6	7	9		EJ.m
	15	22	29	37	45	55	5	13	27	39	42	27	50	53	58	4	10	15	22	30	46	51	25	31	54	4	25	49	58	58		EJ.s
	27	26	26	26	26	26	25	25	25	25	24	24	24	24	24	23	23	23	23	23	22	22	22	22	21	21	21	21	20	20		AH.a
	0	48	37	25	13	2	38	26	14	3	50	37	25	12	0	43	35	22	9	7	43	30	17	7	50	21	23	9	56	42		AH.b
⌜ f. 13r Kh	91	92	93	94	95	96	97	98	99	100	101	102	103	104	105	106	107	108	109	110	111	112	113	114	⌜115	116	117	118	119	120		
	6	6	6	6	6	6	6	6	6	6	6	6	6	6	6	6	6	6	6	6	5	5	5	5	5	5	5	5	5	5		EJ.d
	9	9	9	9	9	9	9	9	10	10	9	8	7	6	5	4	3	2	2	1	58	46	54	52	50	47	45	53	40	38		EJ.m
	16	21	25	30	35	45	50	51	1	7	13	20	27	34	30	27	54	59	5	11	58	44	30	15	0	46	30	14	57	40		EJ.s
	20	20	20	19	19	19	19	18	18	18	18	18	18	17	17	17	16	16	16	16	15	15	15	15	15	14	14	14	14	14		AH.a
	29	15	2	48	34	20	7	13	49	21	12	17	45	32	29	6	52	39	26	12	59	47	34	21⌟	8	15	42	20	16	4	⌟ f. 12v Kh (continues)	AH.b

Table XXI-B

(continued)

121	122	123	124	125	126	127	128	129	130	131	132	133	134	135	136	137	138	139	140	141	142	143	144	145	146	147	148	149	150		
5	5	5	5	5	5	5	5	5	5	5	5	5	4	4	4	4	4	4	4	4	4	4	3	3	3	3	3	3	3		EJ.d
35	32	28	25	21	18	15	11	8	4	0	55	51	41	52	37	33	28	23	19	13	8	2	57	53	46	41	35	30	24		EJ.m
10	0	37	15	52	19	51	40	15	48	19	48	16	42	9	35	1	24	47	10	42	11	56	30	3	35	6	36	6	35		EJ.s
13	13	13	13	12	12	12	12	12	12	11	1[illegible]	11	11	11	11	10	10	10	10	10	10	9	9	9	9	9	9	9	8		AH.a
52	40	28	17	53	41	42	29	10	6	56	45	35	24	13	3	52	42	42	21	12	4	52	47	38	30	21	13	5	56		AH.b
śaniśīghraphalaṃ saṃpūrṇaṃ ‖																															
151	152	153	154	155	156	157	158	159	160	161	162	163	164	165	166	167	168	169	170	171	172	173	174	175	176	177	178	179	180		
3	3	3	3	3	2	2	2	2	2	2	2	2	1	1	1	1	1	1	1	1	0	0	0	0	0	0	0	0	0		EJ.d
18	12	5	59	53	40	33	27	21	14	7	0	0	53	47	40	31	26	19	12	05	58	50	43	36	29	21	14	17	0		EJ.m
18	1	42	23	2	11	58	35	12	23	34	4[illegible]	44	54	3	22	21	29	31	45	58	12	56	40	24	7	51	34	17	0		EJ.s
8	8	8	8	8	8	8	8	7	7	7	7	7	7	7	7	7	7	7	7	7	7	7	7	7	7	7	7	7	7		AH.a
49	43	36	30	23	17	10	4	57	51	47	4[illegible]	40	36	33	29	26	19	18	13	13	12	17	9	7	7	7	4	17	0⌟	⌟ f. 13ʳ Kh	AH.b

Table XXI-C

⌜f. 12ᵛ S43 f. 16ᵛ S45	⌜‖ śaniśīghraphalāni ‖																														
	1	2	3	4	5	6	7	8	9	10	11	12	13	14	15	16	17	18	19	20	21	22	23	24	25	26	27	28	29	30	
	0	0	0	0	0	0	0	0	0	0	1	1	1	1	1	1	1	1	1	1	2	2	2	2	2	2	2	2	2	2	EJ.d
	5	11	17	23	29	35	41	46	52	58	4	9	15	21	26	32	38	43	49	55	0	5	11	16	22	27	33	38	44	49	EJ.m
	51	43	35	28	20	13	5	58	51	43	21	58	35	13	51	29	8	47	26	5	30	57	21	47	14	41	9	36	4	34	EJ.s
	5	5	5	5	5	5	5	5	5	5	5	5	5	5	5	5	5	5	5	5	5	5	5	5	5	5	5	5	5	4	DF.m
	52	52	53	52	52	53	53	53	52	38	37	37	38	38	38	39	39	39	39	25	25	26	26	27	27	27	27	28	30	57	DF.s
	132	132	132	132	132	132	132	132	132	132	132	132	132	132	132	132	132	132	132	132	132	132	132	131	131	131	131	131	131	131	AH.a
	58	57	56	55	54	52	51	50	49	48	45	42	39	36	33	30	27	24	21	18	13	8	3	57	52	46	41	36	31	25	AH.b
	49	39	28	18	8	57	47	36	26	16	20	24	28	33	36	40	44	48	52	56	40	23	6	35	17	57	37	19	0	40	AH.c
	31	32	33	34	35	36	37	38	39	40	41	42	43	44	45	46	47	48	49	50	51	52	53	54	55	56	57	58	59	60	
	2	2	3	3	3	3	3	3	3	3	3	3	3	3	4	4	4	4	4	4	4	4	4	4	4	4	4	4	5	5	EJ.d
	54	59	4	9	14	19	24	29	34	39	44	48	53	57	2	6	11	16	20	25	29	33	36	40	44	48	52	56	0	4	EJ.m
	31	29	28	28	28	29	29	31	33	36	8	41	14	48	23	58	34	11	53	57	16	6	56	47	38	32	25	20	14	9	EJ.s
	4	4	5	5	5	5	5	5	5	5	4	4	4	4	4	4	4	4	5	3	3	3	3	3	3	3	3	3	3	3	DF.m
	58	59	0	0	0	0	2	2	3	32	33	33	34	35	35	36	37	42	4	50	50	50	52	51	54	53	55	54	55	6	DF.s
	131	131	131	130	130	130	130	130	130	130	130	129	129	129	129	129	129	129	128	128	128	128	128	128	127	127	127	127	127	127	AH.a
	18	11	4	57	49	42	35	28	21	14	5	56	47	38	29	20	11	2	50	43	33	23	12	2	52	41	31	20	10	0	⌟f. 12ᵛ S43 AH.b
	35	30	24	4	55	46	38	29	20	11	16	4	4	4	4	4	4	4	51	43	26	9	53	36	1	40	18	57	36	0⌟	f. 16ᵛ S45 AH.c
⌜f. 13ʳ S43 f. 17ʳ S45	⌜61	62	63	64	65	66	67	68	69	70	71	72	73	74	75	76	77	78	79	80	81	82	83	84	85	86	87	88	89	90	
	5	5	5	5	5	5	5	5	5	5	5	5	5	5	5	5	5	5	5	5	5	5	6	6	6	6	6	6	6	6	EJ.d
	7	10	13	16	19	22	26	29	32	35	37	39	41	43	45	48	50	52	54	56	57	58	0	1	2	4	5	6	7	9	EJ.m
	15	22	29	37	45	55	5	16	7	39	42	46	50	53	58	4	10	15	22	30	44	59	15	31	48	4	22	39	58	16	EJ.s
	3	3	3	3	3	3	3	3	3	2	2	2	2	2	2	2	2	2	2	1	1	1	1	1	1	1	1	1	1	1	DF.m
	7	7	8	8	10	10	11	11	12	3	4	4	3	5	6	6	5	7	8	14	14	16	16	17	16	18	17	18	18	5	DF.s
	126	126	126	126	126	125	125	125	125	125	124	124	124	124	123	123	123	123	123	122	122	122	122	122	121	121	121	121	120	120	AH.a
	48	36	24	13	1	49	37	26	14	2	49	37	24	12	59	47	34	22	9	56	43	29	16	3	49	36	22	9	55	42	AH.b
	19	39	59	19	39	40	54	18	22	37	55	36	57	28	45	10	35	0	26	35	16	57	38	19	40	14	49	23	45	8	AH.c

(continues)

Table XXI-C

(continued)

	91	92	93	94	95	96	97	98	99	100	101	102	103	104	105	106	107	108	109	110	111	112	113	114	115	116	117	118	119	120
EJ.d	6	6	6	6	6	6	6	6	6	6	6	6	6	6	6	6	6	6	6	6	5	5	5	5	5	5	5	5	5	5
EJ.m	9	9	9	9	9	9	9	9	10	10	9	8	7	6	5	4	3	2	2	1	58	56	54	52	50	47	45	43	40	38
EJ.s	21	25	30	35	40	45	50	55	1	7	13	20	27	34	40	46	54	59	5	11	58	40	30	15	1	46	30	14	57	40
DF.m	0	0	0	0	0	0	0	0	0	0	0	0	0	0	0	0	0	0	0	2	2	2	2	2	2	2	2	2	2	3
DF.s	4	5	5	5	5	5	5	6	6	54	53	53	54	54	54	54	55	54	54	13	14	14	15	15	15	16	16	17	17	20
AH.a	120	120	120	119	119	119	119	118	118	118	118	117	117	117	117	117	116	116	116	116	115	115	115	115	115	114	114	114	114	114
AH.b	28	15	1	47	34	20	6	52	38	25	12	58	45	32	19	5	52	38	25	12	59	56	33	21	8	55	42	29	16	3
AH.c	36	4	32	39	0	21	42	44	58	12	6	43	30	17	4	50	18	58	38	18	22	36	49	3	11	12	18	25	32	39⌟

⌟ f. 12ᵛ S43 f. 17ʳ S45

⌜ f. 13ᵛ S43 f. 17ᵛ S45

⌜‖ śaniśīghraphalāni ‖

	121	122	123	124	125	126	127	128	129	130	131	132	133	134	135	136	137	138	139	140	141	142	143	144	145	146	147	148	149	150
EJ.d	5	5	5	5	5	5	5	5	5	5	5	4	4	4	4	4	4	4	4	4	4	4	4	3	3	3	3	3	3	3
EJ.m	35	32	28	25	21	18	15	11	8	4	0	59	51	46	42	37	33	28	23	19	13	8	2	57	52	46	41	35	30	24
EJ.s	20	0	37	15	52	29	5	40	15	48	19	48	16	43	9	35	1	24	47	10	46	21	56	30	3	35	6	36	6	35
DF.m	3	3	3	3	3	3	3	3	3	3	3	4	4	4	4	4	4	4	4	4	5	5	5	5	5	5	5	5	5	6
DF.s	20	23	23	23	23	24	25	25	25	29	41	32	33	34	34	34	34	34	37	35	25	24	26	27	28	29	30	30	31	17
AH.a	113	113	113	113	113	112	112	112	112	112	111	111	111	111	111	111	110	110	110	110	110	110	109	109	109	109	109	109	109	108
AH.b	51	40	28	16	5	53	42	29	18	6	55	45	34	24	13	3	52	42	31	21	12	4	55	46	38	29	21	12	4	55
AH.c	47	9	31	53	15	18	34	50	6	22	42	16	49	22	55	28	42	9	37	4	38	12	28	57	27	56	26	55	25	35

	151	152	153	154	155	156	157	158	159	160	161	162	163	164	165	166	167	168	169	170	171	172	173	174	175	176	177	178	179	180
EJ.d	3	3	3	2	2	2	2	2	2	2	2	2	2	1	1	1	1	1	1	1	1	0	0	0	0	0	0	0	0	0
EJ.m	18	12	5	59	53	46	40	33	27	21	14	7	0	53	47	40	33	26	19	12	5	58	50	43	36	29	21	14	7	0
EJ.s	18	1	42	23	3	42	21	58	35	12	23	34	44	54	3	12	21	29	36	44	28	12	56	40	24	7	51	34	17	0
DF.m	6	6	6	6	6	6	6	6	6	6	6	6	6	6	6	6	6	6	6	7	7	7	7	7	7	7	7	7	7	0
DF.s	17	19	19	20	21	23	23	23	23	49	49	50	51	51	51	51	52	53	53	16	16	16	16	16	16	16	16	17	17	0
AH.a	108	108	108	108	108	108	108	108	107	107	107	107	107	107	107	107	107	107	107	107	107	107	107	107	107	107	107	107	107	107
AH.b	49	42	36	29	23	16	10	4	57	50	47	43	40	36	32	29	25	21	18	14	13	11	10	8	7	6	4	3	1	0
AH.c	9	42	16	50	23	57	30	4	20	30	13	36	0	23	46	10	33	58	27	43	16	50	23	56	30	3	36	10	43	0⌟

⌟ f. 13ᵛ S43 f. 17ᵛ S45

Table XXI-C: apparatus criticus

‖ śanīśīghraphalāni ‖] ‖ atha śīghrakendrāṃśopariśaniśīghraphalam aṃśādyantarakarṇaḥ gatiphalāni ‖ śaneḥ śīghraphalāni ‖ (f. 12^{v}) S43, śīghrakendrāṃśopariśanem śīghraphalāni ‖ śaneḥ śīghraphalāni ‖ (f. 13^{r}) S43, śīghrakendrāṃśopariśaniśīghraphalam aṃśādi ‖ śaneḥ śīghraphalāni ‖ (f. 13^{v}) S43; śaniśīghraphalāni ‖ (f. 16^{v}), (f. 17^{r}), (f. 17^{v}) S45.

DF.s(5)53 S43 DF.s(6) 52 S43; AH.b(6) 53 S43 AH.c(8) 22 S43 AH.b(13) 29 S43 AH.c(14) 32 S43 AH.b(19) 25 S43 AH.c(20) 53 S43 EJ.s(22) 55 S43 AH.c(25) 16 S43 DF.s(26) 28 S43 AH.c(27) 38 S43 DF.s(28) 38 S43 DF.s(29) 29 S43 EJ.s(30) 33 S43; DF.m(30) 5 S43; DF.s(30) 53 S43 EJ.s(31) 27 S43; DF.m(31) 5 S43; DF.s(31) 2 S43 EJ.s(32) 28 S43; DF.m(32) 5 S43 DF.s(35) 1 S43 DF.m(40) 4 S43 DF.s(44) 34 S43 DF.s(50) 19 S43 AH.c(51) 36 S43 DF.s(53) 51 S43 DF.s(58) 24 S45 AH.b(60) 1 S43 AH.b(62) 39 S43 AH.b(64) 16 S43 AH.b(68) 28 S43; AH.c(68) 8 S43 EJ.s(69) 27 S43 AH.c(72) 26 S43 AH.c(80) 30 S43 DF.s(81)15 S43 AH.c(83) 8 S43 DF.s(88) 19 S43 AH.c(89) 41 S43 DF.m(99) o dha S43; AH.b(99) 35 S43 DF.m(100) o × S43 AH.c(101) 1 S43 AH.c(102) 42 S43 DF.s(106) 52 S43 AH.c(108) 57 S43 EJ.s(112) 44 S43 AH.b(112) 46 S45 DF.s(114) 14 S43 AH.c(119) 33 S43 DF.s(123) 22 S43 AH.b(124) 26 S43 AH.b(127) 41 S43 DF.s(129) 27 S43 DF.m(130) 4 S43 DF.m(131) 4 S43; DF.s(131) 31 S43 EJ.m(132) 55 S43 DF.s(135) 32 S45 DF.s(137) 37 S43 DF.s(138) 37 S43 DF.m(140) 5 S43, DF.s(140) 24 S43 AH.b(141)11 S43 DF.s(142) 25 S43 AH.b(143) 45 S43; AH.c(143) 2 S43 AH.c(145) 37 S43 EJ.s(146) 36 S43 DF.m(149) 6 S45 AH.c(153) 17 S43 DF.s(156) 21 S43 AH.c(160) 50 S43 DF.s(163) 50 S43 DF.s(168) 52 S43; AH.c(168) 56 S43 DF.s(169) 52 S43; AH.c(169) 20 S43 AH.c(173) 27 S43 DF.s(175) 17 S43 EJ.s(177) 50 S43; DF.s(177) 17 S43; AH.c(177) 34 S43 AH.b(178) 2 S43 AH.s(179) 117 S45.

"iti śrībrahmatulyasya grahasādhanārthaṃ sāraṇī saṃ°" in the left margin of f. 17^{v} S45; "sāraṇīsamāpta ‖ saṃvat 1855 varṣe śāke 1720 pravarttamane kārttikavida 11 some" in the right margin of f. 17^{v} S45 • Table XXI-C *om.* SMB.

Table XXII

⌜a f. 8r Kh, f. 1r S43, f. 7v S45, ⌜b f. 1v S43, ⌜c f. 8v Kh

⌜a‖ bhaumamandoccaspaṣṭīkaraṇārthaṃ koṣṭakāḥ ‖

	1	2	3	4	5	6	7	8	9	10	11	12	13	14	15	⌜b16	17	⌜c18	19	20	21	22	23	24	25	26	27	28	29	30	
MP.d	0	0	0	0	0	0	1	1	1	1	1	1	1	2	2	2	2	2	2	3	3	3	3	3	3	3	4	4	4	4	
MP.m	9	18	27	36	45	54	3	12	21	30	39	48	57	6	15	24	33	42	51	0	9	18	27	36	45	54	3	12	21	30	
MP.s	0	0	0	0	0	0	0	0	0	0	0	0	0	0	0	0	0⌟	0	0	0	0	0	0	0	0	0	0	0	0	0	⌟ f. 8r Kh
DF.m	9	9	9	9	9	9	9	9	9	9	9	9	9	9	9	9	9	9	9	9	9	9	9	9	9	9	9	9	9	9	
DF.s	0	0	0	0	0	0	0	0	0	0	0	0	0	0	0⌟	0	0	0	0	0	0	0	0	0	0	0	0	0	0	0⌟	⌟ f. 1r S43
	31	32	33	34	35	36	37	38	39	40	41	42	43	44	45	46	47	48	49	50	51	52	53	54	55	56	57	58	59	60	
MP.d	4	4	4	5	5	5	5	5	5	6	6	6	6	6	6	6	7	7	7	7	7	7	7	8	8	8	8	8	8	9	
MP.m	39	48	57	6	15	24	33	42	51	0	9	18	27	36	45	14	3	12	21	30	29	48	57	6	14	24	33	42	51	0	
MP.s	0	0	0	0	0	0	0	0	0	0	0	0	0	0	0	0	0	0	0	0	0	0	0	0	0	0	0	0	0	0	
DF.m	9	9	9	9	9	9	9	9	9	9	9	9	9	9	9																⌟ f. 1v S43
DF.s	0	0	0	0	0	0	0	0	0	0	0	0	0	0	0⌟																f. 7v S45
	61	62	63	64	65	66																									
MP.d	9	9	9	9	9	9																									
MP.m	9	18	27	36	45	54																									
MP.s	0	0	0	0	0	0⌟																									⌟ f. 8v Kh

‖ bhaumamandoccasya °...° koṣṭakāḥ] ‖ atha bhaumamandoccaspaṣṭīkaraṇaṃ koṣṭakāḥ ‖ (f. 8r) Kh; ‖ bhaumasya mandoccaspaṣṭārthaṃ koṣṭakāḥ ‖ (f. 1r) S43.

Table XXII *om.* B • DF.#(1)–DF.#(66) *om.* Kh • MP.#(46–66) *om.* S43; "have bhaumanu mandocca spaṣṭa karavāne kāje koṣṭaka chete joyāno prakāra ‖ bhaumāśu kendre padayātagamyeti ‖ svadeśī madhyama bhaumane svadeśī madhyama sūryamāṃthī kāḍī iśekarahete'śukendrathāite aśukendranu padayākījepachegamyakījetepadayāta anegamyamadhyeje alpatehanā aṃśakījete aṃśa upare jo i a i koṣṭakanu phalasantaraśetaghna hāi ‖" vernacular paratext in the left margin of f. 1r S43; "tekoṣṭakanuphalaspaṣṭakarāmadhyamamandoccabhaumanāṃ māṃjo aśukendrakarkādika hoyato ṛṇakīje makarādika hoyato dhanakījete bhaumanu spaṣṭamandoccathāi ‖" vernacular paratext at the beginning of f. 1v S43 • MP.#(46–66) *om.* S45 • Table XXII *om.* SMB.

Table XXIII-A (row: raviḥ) and Table XXIV-A (row: candra)

⌜ f. 1r S43

⌜‖ madhyamaṃ ravisāyanaṃ dvighnaṃ ca vidhāya bhujaḥ kāryas tadaṃśopariraver udayāntaraṃ vikalātmakaṃ tad adhaś candrasya kalādikaṃ sāyane madhyamaravau yugmapadasthe dhanaṃ ojapadasthe ṛṇaṃ ‖ sāyanasūryasya dvighnasya dorjyā śarahṛd viliptā bhānoḥ vidhoḥ kvakṣi 21 hṛtā kalāsphuti udayāntaraṃ kāryam ‖

'ṃśā	1	2	3	4	5	6	7	8	9	10	11	12	13	14	15	16	17	18	19	20	21	22	23	24	25	26	27	28	29	30
raviḥ	0	1	1	1	2	2	3	3	4	4	4	5	5	6	6	6	7	7	8	8	8	9	9	9	10	10	10	11	11	12
candra	0	0	0	0	0	0	0	0	0	1	1	1	1	1	1	1	1	1	1	1	2	2	2	2	2	2	2	2	2	2
	6	12	18	24	30	36	42	48	54	0	6	12	17	23	28	34	40	45	51	57	2	5	11	17	22	28	34	40	46	51
'ṃśā	31	32	33	34	35	36	37	38	39	40	41	42	43	44	45	46	47	48	49	50	51	52	53	54	55	56	57	58	59	60
raviḥ	12	12	13	13	13	14	14	14	15	15	15	16	16	16	16	17	17	17	18	18	18	18	19	19	19	19	20	20	20	20
candraḥ	2	3	3	3	3	3	3	3	3	3	3	3	3	3	4	4	4	4	4	4	4	4	4	4	4	4	4	4	4	4
	54	0	6	8	14	20	26	29	34	40	42	48	51	57	0	6	9	14	17	21	26	29	31	34	40	42	46	49	51	57
'ṃśā	61	62	63	64	65	66	67	68	69	70	71	72	73	74	75	76	77	78	79	80	81	82	83	84	85	86	87	88	89	90
raviḥ	21	21	21	21	21	21	22	22	22	22	22	22	22	23	23	23	23	23	23	23	23	23	23	23	23	23	23	23	23	24
candraḥ	5	5	5	5	5	5	5	5	5	5	5	5	5	5	5	5	5	5	5	5	5	5	5	5	5	5	5	5	5	5
	0	0	2	6	9	15	14	17	20	22	22	26	26	29	29	31	31	34	34	38	38	38	38	38	40	40	40	40	40	42⌟

⌞ f. 1r S43

Table XXIII-B (row: rave) and Table XXIV-B (row: vidhau)

⌜ f. 7v S45

⌜‖ ɛtha ravicandrayor udayāntaraṃ koṣṭakāḥ ‖ dvighnaṃ bhujāṃśopari ‖

	2	4	6	8	10	12	14	16	18	20	22	24	26	28	30	32	34	36	38	40	42	44	46	48	50	52	54	56	58	60	
rave	0	1	2	3	4	5	5	6	7	8	8	9	10	11	12	12	13	14	14	15	16	16	17	18	18	18	19	19	20	20	
	50	41	31	22	12	0	48	36	24	12	57	43	29	14	0	41	22	2	43	24	0	36	12	48	24	53	22	50	19	48	vikalā
vidhau	0	0	0	0	1	1	1	1	1	1	2	2	2	2	2	3	3	3	3	3	3	3	4	4	4	4	4	4	4	4	kalā
	12	24	36	48	0	11	22	34	45	57	8	15	29	40	51	1	10	28	30	40	48	57	5	14	22	29	36	43	50	57	vikalā
	0	0	0	0	0	26	51	17	43	9	0	52	43	35	26	9	15	34	17	0	34	8	53	17	51	43	34	26	17	8	

	62	64	66	68	70	72	74	76	78	80	82	84	86	88	90
rave	21	21	21	22	22	22	23	23	23	23	23	23	23	23	24
	10	31	53	14	36	48	0	12	24	36	41	45	50	55	0
vidhau	5	5	5	5	5	5	5	5	5	5	5	5	5	5	5
	2	7	12	17	22	25	28	31	34	37	38	39	40	41	42
	17	25	34	43	51	43	34	26	17	7	17	26	34	43	51⌟

⌟ f. 7v S45

Table XXIII-C

⌜ f. 5^{v} SMB

⌜‖ atha dvighnasāyanaravibhujāśopari yātaphalaṃ kalādi | ojapade ṛṇaṃ | yugmapade dhanaṃ ‖																													
1	2	3	4	5	6	7	8	9	10	11	12	13	14	15	16	17	18	19	20	21	22	23	24	25	26	27	28	29	30
0	0	0	0	0	0	0	0	0	0	0	0	0	0	0	0	0	0	0	0	0	0	0	0	0	0	0	0	0	0
0	1	2	2	2	2	3	3	3	4	4	5	5	6	6	6	7	7	8	8	8	8	9	9	10	10	10	11	12	12
31	32	33	34	35	36	37	38	39	40	41	42	43	44	45	46	47	48	49	50	51	52	53	54	55	56	57	58	59	60
0	0	0	0	0	0	0	0	0	0	0	0	0	0	0	0	0	0	0	0	0	0	0	0	0	0	0	0	0	0
12	12	13	13	13	14	14	14	15	15	16	16	16	16	16	17	17	18	18	18	18	18	19	19	19	20	20	20	20	20
61	62	63	64	65	66	67	68	69	70	71	72	73	74	75	76	77	78	79	80	81	82	83	84	85	86	87	88	89	90
0	0	0	0	0	0	0	0	0	0	0	0	0	0	0	0	0	0	0	0	0	0	0	0	0	0	0	0	0	0
20	21	21	21	21	21	22	22	22	22	22	22	22	23	23	23	23	23	23	23	23	23	23	23	23	23	23	23	23	24⌟

⌞ f. 5^{v} SMB

Table XXIV-C

⌜ f. 5^{v} SMB

⌜‖ atha candrasya yāntaphalaṃ vikalādiravivaddhanarṇaṃ ‖																													
1	2	3	4	5	6	7	8	9	10	11	12	13	14	15	16	17	18	19	20	21	22	23	24	25	26	27	28	29	30
0	0	0	0	0	0	0	0	0	1	1	1	1	1	1	1	1	1	1	1	2	2	2	2	2	2	2	2	2	2
6	12	18	24	30	36	42	48	54	0	5	11	17	23	28	34	40	45	51	57	2	8	14	19	25	29	35	40	46	51
31	32	33	34	35	36	37	38	39	40	41	42	43	44	45	46	47	48	49	50	51	52	53	54	55	56	57	58	59	60
2	3	3	3	3	3	3	3	3	3	3	3	3	3	4	4	4	4	4	4	4	4	4	4	4	4	4	4	4	4
56	1	6	11	15	20	25	30	35	40	44	48	53	57	1	5	10	14	18	23	26	29	33	36	40	43	47	50	51	57
61	62	63	64	65	66	67	68	69	70	71	72	73	74	75	76	77	78	79	80	81	82	83	84	85	86	87	88	89	90
4	5	5	5	5	5	5	5	5	5	5	5	5	5	5	5	5	5	5	5	5	5	5	5	5	5	5	5	5	5
59	2	5	8	10	13	15	18	20	23	24	26	27	29	30	31	33	34	35	37	38	38	39	39	40	41	41	42	42	43⌟

⌞ f. 5^{v} SMB

Table XXV FROM MS KH 5424 (A)

⸢f. 1ᵛ Kh																
	⸢1470	1500	1530	1560	1590	1620	1650	1680	1710	1740	śākāyam ‖					
	133331	144287	155243	166199	177155	188111	199067	210023	220979	231935	karṇāharganāḥ ‖					
	1	2	3	4	5	6	7	8	9	10	11	12	13	14	15	
	384	739	1092	1477	1831	2215	2570	2924	3308	3662	4046	4400	4755	5139	5493	
	16	17	18	19	20	21	22	23	24	25	26	27	28	29	30⸥	⸥ f. 1ᵛ Kh

Table XXVI FROM MS KH 5424 (A)

⸢f. 1ᵛ Kh	⸢māsapraveśa kṣepāṃkapatraṃ‖												
	1	2	3	4	5	6	7	8	9	10	11	12	māsāḥ
	3	31	31	31	31	30	29	29	29	29	29	30	vāra
	56	26	37	28	2	27	53	28	59	26	48	26	ghaṭī
	55	21	8	18	2	0	10	59	34	47	51	27	pala·

dinapraveśakṣepakapatram													
1	2	3	4	5	6	7	8	9	10	11	12	dinántarāṇi	
1	1	1	1	1	1	1	0	0	0	0	1	vārāḥ	
1	1	3	2	2	0	59	58	58	58	59	0	ghaṭī·	
53	50	14	26	4	54	46	57	39	53	37	42	palāni	
50	42	16	36	4	0	20	58	8	34	42	0	karāṇi ‖⸥	⸥ f. 1ᵛ Kh

"atha brahmatulyoktagrahāṇāṃ spaṣṭīkaraṇe koṣṭakānusāreṇa sūtraṃ pratipādayati kendrasya doḥ kāryāḥ yāvatyaṃśā bhavanti tāvanti maḥ koṣṭakaḥ karṣaṇīyaḥ etāvatādūre likhanīyaḥ'gre koṣṭakastanmadhe pātanīyaḥ kalāvikalābhavanti tadgaumūtrikānyāsena guṇyate bujāṃśādhaḥ kalāvikalāguṇite ṣaṣṭyā'dhṛte dhanaṃ spaṣṭamandaphalaṃ bhavati· svamṛṇaṃ kāryaṃ gṛhe budhādayo grahāḥ saumyāḥ karma catuṣṭayena spaṣṭā bhavanti· mandārdhaśīghrārdhasakalamandārdha-kalasakalaśīghrārdhaphaladattesati spaṣṭā bhavanti paraṃtvayaṃ viśeṣa bhauma ekonatriṃśat karmayāvat spaṣṭona bhavantīti viśeṣaḥ | ahargaṇe 30 bhāgeṃdījeṃ śeṣa raha i te dinakoṣṭakalabdhaniṃ 12 śeṣa raha i te māsakoṣṭakalabdhaniṃ 20 bhāgedīja iśeṣaṃ varṣakoṣṭaka· ekoṣṭaka 4 ekatrakījeṃ e tala i etala i madhyamalaṃkodaya nāthāi" written at the top of f. 1ᵛ Kh; "śakānā'hargaṇamāṃhi antarāṣu varṣanā ahargaṇajorī etivāra ekāguṇāvadi amāvasyāno'hargaṇā[-x-] va i te madhye | caitrādika gatamāsatītithikījaratemāhi thīvimāse ekekītithikāḍhīya i śeṣarahitevarṣāhargaṇamadhye jomīya i· tadinato'hargaṇabrahmatulyano [-x-]va i ‖" written to the right of f. 1ʳ Kh; "‖ varṣapraveśavārādi madhye prathamadinapraveśakṣepāṃka yute dvitīyadinapraveśavārādi· ima 12 dina lage pache ivalī prathamadivasanākṣepakamelīya i· imā 365 dinalage karavu iti dinapraveśaḥ ‖" written to the right of f. 1ᵛ Kh; "je dina'varṣapraveśa hu i te dinanā vāraghaṭīpalamāṃḍī prathamamāsanākṣepāṃka melī i vārādikane 7 bhāgadījeṃ dvitīyamāsanuṃvārādi hu i· 1 va· 12 māsāḥ ‖" written at the bottom of f. 1ᵛ Kh.

Tables XXV–XXVI only found in Kh; "| $\overline{\text{ta}}$ ×" (correction/insertion?) in right margin on f. 1ᵛ Kh.

Table XXVII

⌜ f. 8v Kh	⌜‖ ravyādayo ⌜grahāmandoccāḥ ‖								deśāntaram idam aṃśādi																			
	ra	maṃ	bu	vṛ	śu	śa	gra		ra	caṃ	maṃ	bu	vṛ	śu	śa	u	pā		ra	caṃ	maṃ	bu	vṛ	śu	śa	u	pā	
	2	2	7	5	2	7	rā		0	0	0	0	0	0	0	0	0		0	0	1	11	3	4	1	0	0	
	18	8	15	12	28	28	'ṃ		0	5	0	1	1	1	0	0	0		2	15	2	4	10	30	30	0	30	
	0	30	0	30	0	0	ka		44	24	2	32	13	41	2	19	0		0	0	0	0	10	0	0	30	0	
	0	0	0	0	0	0	vi												dha·	×	dha·	dha·	×	×	dha·	dha·	×⌟	⌞ f. 8v Kh

Table XXVII only found in Kh.

Table XXVIII

⌜a f. 13r Kh ⌜b f. 13v Kh

⌜a‖ krāntikalākoṣṭakāḥ ‖

1	2	3	4	5	6	7	8	9	10	11	12	13	14	15	16	17	18	19	20	21	22	23	24	25	26	27	28	29	⌜b30	
14	48	72	75	229	243	267	291	215	239	262	266	309	333	356	372	403	426	449	471	494	517	539	54	584	606	627	649	671	692	SL.m
0	0	49	57	54	49	40	30	18	2	42	17	48	16	27	53	2	5	2	53	36	11	36	55	4	7	54	33	3	22	SL.s
24	23	23	23	23	23	23	23	23	23	23	23	23	23	23	23	23	22	22	22	22	22	22	22	21	21	21	21	21	21	DF.m
0	59	59	57	55	51	50	48	44	40	35	32	38	22	16	9	3	57	53	43	35	25	19	9	59	51	39	30	13⌟	6	⌞ f. 13r Kh DF.s
31	32	33	34	35	36	37	38	39	40	41	42	43	44	45	46	47	48	49	50	51	52	53	54	55	56	57	58	59	60	
697	734	755	775	795	816	838	855	874	894	913	931	950	968	286	102	158	1037	1054	701	86	1102	1118	1233	1247	1262	1276	1290	123	1216	SL.m
21	25	9	40	59	4	55	39	59	6	1	44	2	9	1	34	44	49	55	50	53	38	2	6	50	15	19	0	20	20	SL.s
20	20	20	20	20	19	19	19	19	18	18	18	18	17	17	17	17	16	16	0	0	0	0	0	0	0	0	0	0	0	DF.m
57	44	31	19	5	51	44	20	7	55	43	16	7	52	33	14	1	40	21												DF.s
61	62	63	64	65	66	67	68	69	70	71	72	73	74	75	76	77	78	79	80	81	82	83	84	85	86	87	88	89	90	
1228	41	52	63	75	84	96	1306	15	24	33	41	49	57	64	70	76	89	91	90	139	1403	3	6	1409	11	12	14	14	1415	SL.m
57	11	2	29	35	10	14	27	54	59	17	53	42	5	1	32	35	14	27	14	32	25	50	46	18	22	58	5	47	0⌟	⌞ f. 13v Kh SL.s
0	0																													DF.m
																														DF.s

DF.S(50)–DF.S(62) *om.* f. 13v Kh; DF.#(63)–DF.#(90) *om.* f. 13v Kh; Table of ‘Rising-differences’ (*udayāntara*) correction for the Moon *om.* Kh.

Appendix: Sanskrit Astronomy and the *Karaṇakutūhala*

This appendix collects various supplementary information about the material in this study, much of which will already be familiar to students of Sanskrit *jyotiṣa* or astral sciences.[1]

1 Conventions Used for Sanskrit Text, Numbers and Calendar Reckoning

1.1 *Representation of Sanskrit*

For edited or transcribed Sanskrit text in *nāgarī* script we use the Shobhika font with the typographic conventions described in section 5.1. In quoting text from the *Brahmatulyasāraṇī* and *Karaṇakutūhala* in chapter 3 and Appendix, section 2, we use instead roman transliteration following the International Alphabet of Sanskrit Transliteration (IAST) scheme, to make the identification of Sanskrit technical terms as transparent as possible for all readers. (The transliterated text separates individual words regardless of Sanskrit conventions for sandhi, i.e., orthographic indication of morphophonological changes. Occasionally a word representing a *bhūtasaṃkhyā* number (see Appendix, section 1.2) will be interrupted by the decimal numeral denoting that number, as in the edited text.)

Given the centrality of the Sanskrit didactic verse genre in both composing astronomical works and editing their textual variants, we have accompanied transliterated verses with marginal notes indicating the name of the verse metre used therein (*vide* (Apte, 1890, pp. 1179–1189) for details of Sanskrit metres). Regardless of metre, each verse is laid out with its four *pādas* or quarter-verses on separate lines. We observe sandhi rules across the *pāda* boundary within a half-verse but break sandhi between two half-verses.

Our translations of Sanskrit technical terms mostly accord with those in the glossary of (Montelle and Plofker, 2018, Appendix C), which contains more detailed information on many of them.

1 Summaries of several aspects of the immense subject of *jyotiṣa*, along with bibliographies identifying many important studies of it, can be found in (Ruggles, 2015, Vol. 3, Part XII).

 | DOI:10.1163/9789004432222_008

1.2 *Numerical Notation*

By default, sexagesimal (base-60) place-value numerals obey the convention of separating sexagesimal 'digits' by commas except where a semicolon separates the integer and fractional parts of the number, as in

$$a, b\,; c, d, e \quad = \quad a \times 60 + b \times 1 + \frac{c}{60} + \frac{d}{3600} + \frac{e}{216000}.$$

Celestial longitudes, however, are generally represented in units of zodiacal signs (^{s}) and degrees ($^{\circ}$) with the sexagesimal fractional part of the number denoting arcminutes ($'$), arcseconds ($''$), and so on. Thus, for example,

$$a^{s}, b^{\circ}; c, d, e \quad = \quad (a \times 30 + b)^{\circ}\, c'\, d''\, e'''.$$

Sexagesimal notation is similarly used to denote time intervals in civil days (^{d}) and the base-60 subdivisions of a day. For example,

$$a^{d}; b, c \quad = \quad a \text{ days, } b \textit{ ghaṭikās, } c \textit{ vighaṭikās}.$$

In Sanskrit manuscripts, and sometimes in our transcriptions of manuscript data, sexagesimal numerals are written with their 'digits' separated by vertical lines or *daṇḍas*, as in

$$a\,|\,b\,|\,c\,|\,d\,|\,e.$$

Such numbers generally have no special symbol for denoting units or separating integer and fractional parts, so the power of 60 each 'digit' represents has to be determined from the context.

Numerals within Sanskrit text are usually encoded in the *bhūtasaṃkhyā* or 'word-numeral' system alluded to in section 5.1. For a detailed explanation of this system and its standard vocabulary, *vide* (HAMSI, 2017).

1.3 *Indian Calendar Systems*

The complex luni-solar calendars of Sanskrit astronomy, which synchronize solar years and synodic (lunar) months by occasional intercalation of a 'leap month' (*adhimāsa*), are described in detail in, for instance, (Plofker and Knudsen, 2011), (Yano, 2003), and (Sewell and Dikshit, 1896). We have relied mostly on (Yano, 2004) for date conversions, in which Common Era dates before 15 October 1582 are given in the Julian calendar and subsequent ones in the

Gregorian. Due to the possibility of individual variation in local versions of calendars, the identification of dates is not always certain.

1.3.1 Eras and Their Conversion

The most commonly used solar-year calendar eras in Indian astronomy are the Saṃvat, beginning in 57 BCE, and the Śaka, beginning in 78 CE. Thus a year number in one of these eras is roughly converted to its Common Era equivalent by subtracting 57 or adding 78, respectively. However, since Indian calendars typically put the beginning of the year at a point near an equinox (see below) rather than the Julian or Gregorian January 1, a stated Śaka or Saṃvat year number will span parts of two consecutive Common Era years.

1.3.2 Year Reckoning: Meṣasaṅkrānti and Caitraśuklapratipad

The year in most Indian calendars begins either with Meṣasaṅkrānti, the Sun's entry into the first sidereal zodiac sign Meṣa (Aries), or with Caitraśuklapratipad, the start of the synodic month Caitra. Both of these events occur not far from the vernal equinox when the Sun reaches the beginning of the tropical zodiac. Some calendar systems, on the other hand, place the start of the year at the beginning of the month Kārttika, falling near the autumnal equinox. In all cases, the observer's presumed position defaults to the traditional Indian prime meridian passing through Laṅkā and Ujjayinī; *vide* Appendix, section 2.3.

1.3.3 Month Reckoning: *amānta* and *pūrṇimānta*

Depending on the individual calendar system, the start of a synodic (lunar) or calendar month is assigned either to the moment of conjunction of Sun and Moon (new moon, *amānta*) or else to their opposition (full moon, *pūrṇimānta*). A solar (*saura*) month is one-twelfth of a sidereal year, or the average time that the Sun spends in each zodiacal sign.

1.3.4 Bright and Dark Fortnights: *śuklapakṣa* and *kṛṣṇapakṣa*

A synodic month consists of a waxing or 'bright' half (*śuklapakṣa*) from new moon to full moon, and a waning or 'dark' half (*kṛṣṇapakṣa*) from full moon to new moon. Thus the *śuklapakṣa* is the first half of a month in an *amānta* system, but the second half of a *pūrṇimānta* month.

1.3.5 Days and Their Subdivisions, and *tithis*

A mean civil day is considered to be the average time between two consecutive sunrises or two consecutive midnights, depending on how a given astronomical *pakṣa* or school (not to be confused with the abovementioned half-month *pakṣas*) reckons time. For more precise timekeeping, one day is divided into 60 equal *ghaṭikās*, each of which comprises 60 *vighaṭikās* or *palas*, which in turn are divided into 60 *vipalas*. A mean *tithi* is one-thirtieth of a mean synodic month.

2 Computational Formulae in the *Karaṇakutūhala*

In this section we explain mathematical and textual justifications for the formulae prescribed by Bhāskara in the *Karaṇakutūhala* and adapted in the *Brahmatulyasāraṇī* to tabular equivalents. For the most part, these formulae are simplified approximations to those in Bhāskara's comprehensive spherical astronomy treatise *Siddhāntaśiromaṇi*. The quoted *Karaṇakutūhala* verses are taken from the editions in (Mishra, 1991) and (Plofker, n.d.), with silent emendation of typographical errors and other trivial changes such as doubling versus non-doubling of consonants after 'r'.

2.1 *Computation of the Time since Epoch or* ahargaṇa

In the Sanskrit tradition, the beginning of astronomical calculations for a given point in time is usually the computation of the *ahargaṇa* or elapsed time *c* since the epoch of the text. As noted in section 2.1, the epoch of the *Karaṇakutūhala* is sunrise on Thursday 1 Caitra of Śaka 1105, or 24 February 1183. The determination of *c* is prescribed as follows in *Karaṇakutūhala* 1.2–3 (Mishra, 1991, p. 2):

śakaḥ pañcadikcandrahīno 'rkanighno
madhor yāta māsānvito 'dho dvinighnāt ||
rasāṅgānvitāt svābhrakhāṅkāṃśahīnāc
charāṅgair avāptādhimāsair yugūrdhvaḥ || 1.2 || *bhujaṅgaprayāta*

kharāmāhato yāta tithyanvito 'dhas
triyuktāt kharāmābhraśailāṃśayuktāt ||
yugāṅgair avāptāvamonas tadūrdhvo
bhavej jīvavārādiko 'hargaṇo 'yam || 1.3 || *bhujaṅgaprayāta*

VERSES 1.2–3

The [current] Śaka [year], diminished by 1105, multiplied by 12, increased by the months elapsed since Madhu [Caitra], [is placed] below. [This result is] added to the intercalary months [obtained] by dividing by 65 [that same result] multiplied by 2, increased by 66, and diminished by

> 1/900 of itself, [and placed] above. That [result placed] above, multiplied by 30, increased by elapsed *tithis*, is diminished by the omitted *tithis* [obtained] by dividing by 64 [that same result] which, [placed] below, [was] increased by 3 and increased by 1/703 of itself. This should be the *ahargaṇa*, beginning with Thursday.

In other words, knowing the current Śaka year y_0, one first finds the accumulated integer solar years y since epoch from the relation

$$y = y_0 - 1105.$$

The accumulated solar (*saura*) months or twelfths of a year, denoted m_S, are given by

$$m_S = 12y + m_0,$$

where m_0 is the number of calendar months elapsed since 1 Caitra of the current year. (In theory, these m_0 synodic months should be converted to *saura* months before being included in m_S.)

Since a synodic or calendar month is shorter than one-twelfth of a year, the total number m of accumulated synodic months since epoch must be greater than m_S. The difference between them is the number a of extra or intercalary months that occur in this time interval, which according to Bhāskara's rule implies

$$m = m_S + a = m_S + \frac{2m_S + 66 - \dfrac{2m_S + 66}{900}}{65}.$$

The number t of accumulated *tithis*, or thirtieths of a mean synodic month, since epoch is then given by

$$t = 30m + t_0$$

where t_0 is the number of elapsed *tithis* since the beginning of the current month. But since an actual (civil) day is longer than one-thirtieth of a month, the number of days since epoch, or *ahargaṇa*, must be smaller than t. Their difference is the number u of so-called 'omitted' *tithis* in this time period. So the *ahargaṇa* c is finally arrived at by computing

$$c = t - u = t - \frac{t + 3 + \dfrac{t+3}{703}}{64}.$$

Our reconstruction of how Bhāskara derived his prescribed expressions for a and u from the canonical numbers of celestial cycles assigned by the Brāhmapakṣa to the *kalpa* or lifetime of the universe is explained in the following paragraphs.[2]

The *kalpa* consists of $Y = 4320000000$ years beginning with the simultaneous occurrence of the start of the calendar (synodic) month Caitra at new moon, and the start of the solar year with the Sun's entrance into the (sidereal) zodiacal sign Meṣa or Aries, called Meṣasaṅkrānti. The ratio of the number A of intercalary months in Y years to the corresponding number M_S of *saura* months is

$$\frac{A}{M_S} = \frac{A}{12Y} = \frac{1593300000}{51840000000} = \frac{15933}{518400}.$$

Hence the proportional number a of intercalary months in m_S *saura* months can be represented by

$$a = m_S \cdot \frac{A}{M_S} = m_S \cdot \frac{15933}{518400} \approx m_S \cdot \frac{1}{32 + \dfrac{1}{2}} = \frac{2m_S}{65}.$$

But Bhāskara's rule specifies instead $a \approx \frac{2m_S + 66 - \frac{2m_S+66}{900}}{65}$, suggesting that he tweaked this approximation algebraically to improve its accuracy. We can reproduce his modification by postulating a factor, say h, such that

$$h\left(\frac{2m_S}{65}\right) = m_S \cdot \frac{15933}{518400}$$

or

$$h = \frac{65}{2} \cdot \frac{15933}{518400} = \frac{1035645}{1036800} = \frac{1035645}{900 \cdot 1152} = \frac{1}{900} \cdot \left(898 + \frac{1149}{1152}\right) \approx \frac{899}{900},$$

yielding

2 For a fuller description of the characteristics and evolution of the *kalpa* and other standard cycles in Indian cosmology, *vide* especially (González-Reimann, 2009, pp. 415–422).

$$a \approx \frac{899}{900} \cdot \frac{2m_S}{65},$$

which is still not exactly the form of a stated by Bhāskara. We can explain the discrepancy by noting that this expression for a will account for the intercalary months amassed since the epoch date, but not the fraction of an intercalary month that had accumulated in the time before the epoch. That quantity can be found in the same way from ratios of the *kalpa* parameters, as follows.

The integer number of solar years from Meṣasaṅkrānti at the start of the *kalpa* to mean Meṣasaṅkrānti of the epoch year mostly consists of the (4567 × 432000) years up to the start of the present 'Kali Era' or Kaliyuga beginning on 18 February 3102 BCE. To this is added 4284 elapsed years of the Kaliyuga, yielding 1972948284 years in all; *vide* (González-Reimann, 2009, pp. 418–419). The corresponding number of synodic months is derived by proportion from the number M of synodic months in a *kalpa*, as follows:

$$\frac{M}{Y} \cdot 1972948284 = \frac{53433300000}{4320000000} \cdot 1972948284 = 24403041098 + \frac{9}{400}.$$

But the actual epoch date of 1 Caitra or 24 February fell a little more than one synodic month before the end of this time period at mean Meṣasaṅkrānti; so the elapsed integer synodic months M_E at epoch must be 24403041097.[3] The corresponding number a_E of intercalary months accumulated over that period is

$$a_E = \frac{A}{M} \cdot M_E = \frac{1593300000}{53433300000} \cdot 24403041097 = 727661689 + \frac{176688}{178111}.$$

The fractional intercalary month at epoch is therefore 176688/178111. The time it takes to accumulate this fraction of an intercalary month is x *saura* months, where x is given by

$$x = \frac{M_S}{A} \cdot \frac{176688}{178111} = 32 + \frac{261365728}{945947521}.$$

3 This deduction is confirmed by the fact that no other plausible value for M_E produces a value of the constant x (described below) that is anywhere near the 33 of Bhāskara's rule.

So if the quantity m_S in the formula for a is increased by x, the resulting value of a will include the fractional part of a_E, as desired. It appears that for simplicity's sake, Bhāskara rounded x up to 33 (although it would be slightly more accurate to round it down to 32):

$$a \approx \frac{899}{900} \cdot \frac{2(m_S + 33)}{65} = \frac{899}{900} \cdot \frac{2m_S + 66}{65}.$$

The formula for the number u of omitted *tithis* accumulated since epoch is similarly derived by means of the *kalpa* parameters T and U, the numbers of total *tithis* and omitted *tithis* in a *kalpa*, respectively:

$$\frac{U}{T} = \frac{U}{30M} = \frac{25082550000}{1602999000000},$$

so

$$u = t \cdot \frac{U}{T} \approx t \cdot \frac{1}{63 + \cfrac{1}{1 + \cfrac{1}{10}}} = t \cdot \frac{11}{703} = t \cdot \frac{1}{703} \cdot \frac{704}{64} = \frac{t + \cfrac{t}{703}}{64}.$$

Like the first approximation for a discussed above, this approximation[4] $u \approx t \cdot \frac{11}{703}$ differs slightly from the u specified in the verse, because it does not account for the fractional amount of a time-unit already accumulated at epoch. That amount is the fractional part of the number u_E of the total omitted *tithis* at epoch, found from the *kalpa* ratios as follows:

$$u_E = \frac{U}{M} \cdot M_E = \frac{25082550000}{53433300000} \cdot 24403041097 = 11455225458 + \frac{17373}{356222}.$$

The fractional omitted *tithi* at epoch, 17373/356222, can be accumulated in z *tithis* where z is found from the ratio of T to U:

4 The expression $t \cdot \frac{11}{703}$ is used to calculate u in Sanskrit astronomical texts as early as the *Romakasiddhānta* of Varāhamihira at the start of the sixth century (Neugebauer and Pingree, 1970–1971, I, 12, 29, and II, 8–9); Bhāskara probably recast it in this slightly more complicated form to parallel the form of the expression for intercalary months.

$$z = \frac{T}{U} \cdot \frac{17373}{356222} = 3 + \frac{6513}{55739}.$$

These z *tithis*, rounded down to 3, have evidently been added in with the t *tithis* elapsed since epoch to give a value for u that includes the fractional part of u_E:

$$u \approx t + 3 + \frac{t + \frac{t}{703}}{64},$$

as Bhāskara's rule states.

2.2 *Planetary Epoch Mean Longitudes and Mean Longitude Increments since Epoch*

The concept of mean celestial motions in Indian geocentric astronomy embraces the Sun, the Moon, Mars, Jupiter, Saturn, and four notional objects determined by orbital parameters: namely, the ascending node of the Moon's orbit, the apogee of the Moon's epicycle, and the apogees of the so-called *śīghra*-epicycles of Mercury and Venus. The former two objects account for the relatively rapid cycles of nodal precession and apsidal precession, respectively, when computing the Moon's position.[5] The latter two are used instead of the actual celestial bodies of Mercury and Venus, since the mean motions of those bodies are considered identical to that of the Sun.

Each of these celestial objects, which for convenience we will here lump together under the term 'planets', completes some integer number K of 360° revolutions about the earth during the $Y = 4320000000$ years of a *kalpa*. So its mean ecliptic longitude $\bar{\lambda}$ after any integer number Y' of solar years since the start of the *kalpa* will be the total number of degrees of arc it has traversed in that time, modulo 360°:

5 The lunar node is considered to revolve from east to west, in the opposite direction to all the other planetary revolutions. We have indicated this by applying a negative sign in parentheses to the parameters for the node in Schemas 14, 15, and 16. Bhāskara in this part of the *Karaṇakutūhala* treats the nodal parameters as positive, ultimately producing a value for the node's mean longitude that is the 360-complement or 'explement' of its actual position. He resolves this apparent inconsistency when the longitudinal elongation between the node and another celestial body is required (in subsequent material not dealt with in the *Brahmatulyasāraṇī*) by taking the *sum* of the two bodies' longitudes rather than their difference. *Vide*, e.g., *Karaṇakutūhala* 4.5 (Mishra, 1991, p. 51) and *Siddhāntaśiromaṇi* 4.3 (Śāstrī, 1989, pp. 109–111).

$$\bar{\lambda} = \left(\frac{360\,K}{Y} \cdot Y'\right) \bmod 360°.$$

Because the numbers K and Y are so large, this is not a very practical formula for computing planetary mean longitudes at some arbitrary date within a given calendar year. In an astronomical handbook with a specified epoch such as the *Karaṇakutūhala*, a planet's mean longitude at the end of some elapsed *ahargaṇa* c since the epoch is calculated instead from the planet's epoch mean longitude $\bar{\lambda}_E$ supplied by the text, and its average angular velocity or 'mean daily motion' $\bar{\nu}$ in degrees per day:

$$\bar{\lambda} = \left(\bar{\lambda}_E + \bar{\nu} \cdot c\right) \bmod 360°.$$

The mean velocity $\bar{\nu}$ is ultimately derived from the *kalpa*-parameter ratio

$$\bar{\nu} = \frac{360\,K}{C}$$

where $C = T - U$ is the total number of civil days in a *kalpa*.

2.2.1 Epoch Mean Longitudes

Bhāskara begins his discussion of mean longitudes in *Karaṇakutūhala* 1.4–6 (Mishra, 1991, p. 4) by specifying the planets' 'offsets' or epoch mean longitudes $\bar{\lambda}_E$ at the start of calendar Śaka 1105:

diśo go yamā viśvatulyāḥ khamarke
vidhau khendavo 'ṅkāśvinaḥ pañcakhākṣāḥ ||
vidhūcce 'bdhayo 'kṣendavo 'rkā navākṣā
navātyaṣṭitattvā grahāś candrapāte || 1.4 || *bhujaṅgaprayāta*

kuje 'śvāḥ kudasrā jināḥ kvakṣitulyā
budhe dvau kunetrāṇi śakrāḥ kharāmāḥ ||
gurau kṣepako dvau kṛtāḥ khaṅkubāṇāḥ
site 'ṣṭau dhṛtir mārgaṇāḥ pañcabāṇāḥ || 1.5 || *bhujaṅgaprayāta*

yugāny agnayas tryabdhayaḥ śailacandrāḥ
śanau ceti rāśyādinā kṣepakeṇa ||
dyupiṇḍotthakheṭo yutaḥ svena madhyo
bhaved udgame 'rkasya laṅkānagaryām || 1.6 || *bhujaṅgaprayāta*

Verses 1.4–6

> 10 [zodiacal signs], 29 [degrees], [minutes] equal to 13, 0 [seconds] in the case of the Sun; in the case of the Moon, 10, 29, 5, 50; in the case of the Moon's apogee, 4, 15, 12, 59; and 9, 17, 25, 9 in the case of the Moon's node. In the case of Mars, 7, 21, 24, and [seconds] equal to 21; in the case of Mercury, 2, 21, 14, 30; in the case of Jupiter, the offset is 2, 4, 0, 51; in the case of Venus, 8, 18, 5, 55; and 4, 3, 43, 17 in the case of Saturn. [The longitude of] the planet derived from the *ahargaṇa* is to be increased by its own offset thus [stated], beginning with signs. [That] is the mean [longitude] at sunrise at the city of Laṅkā [at the prime meridian.]

These epoch mean longitude values are transcribed in column 2 of Schema 14. Since the time interval between the beginning of the *kalpa* and the epoch is not an integer number of solar years, their derivation requires some further explanation (which Bhāskara does not provide). We reconstruct its basic rationale as follows:

1. A proportion involving the *kalpa* length *Y*, a planet's number *K* of revolutions in a *kalpa* (listed in column 3 of Schema 14), and the 1972944000 solar years from the beginning of the *kalpa* to the start of the Kaliyuga gives the planet's mean longitude at the start of the Kaliyuga.
2. The planet's mean longitude increment in the 4284 solar years from the start of the Kaliyuga to mean Meṣasaṅkrānti of the epoch year Śaka 1105 is likewise computed by proportion from *K* and *Y*.
3. This increment is then corrected by 4284 times a small factor known as the planet's yearly *bīja* (*vide Siddhāntaśiromaṇi* 1.6.7–8 (Śāstrī, 1989, p. 38)), a correction with a period of 12000 years that is zero for every planet at the beginning of the Kaliyuga; for intervals within recent historical time, therefore, it is essentially a secular correction. This correction is here called the Brāhma-*bīja* to reflect its use by earlier astronomers in the Brāhmapakṣa, and to avoid confusion with a similar correction known as the *rāmabīja*, discussed below.[6] Its values for the planets, in units of arcminutes per year, are listed in column 4 of Schema 14. The resulting net increment is then added (modulo 360°) to the mean longitude at the

6 See (Pingree, 1996) for a discussion of the use of *bīja*-corrections in Indian astronomy, referencing also a different hypothesis of R. Billard; in it, Table 1 lists these planetary Brāhma-*bījas*, and Table 5 reconstructs their effects on planetary data in the *Karaṇakutūhala*. Note that our reconstruction of the *Karaṇakutūhala*'s use of the Brāhma-*bījas* in the following exposition differs slightly from Pingree's.

start of the Kaliyuga, giving the mean longitude after 1972948284 years since creation, i.e., at mean Meṣasaṅkrānti of Śaka 1105.

4. As we saw in the discussion of verses 2–3, however, that moment does not coincide precisely with the (mean) beginning of the calendar year Śaka 1105 at the start of a mean synodic month. Not only is there an accumulated fraction of 9/400 synodic months or 27/40 *tithis* between Meṣasaṅkrānti and the end of the synodic month preceding it, but that preceding month itself is an intercalary one inserted at the beginning of 1105.

   So to find a planet's mean longitude at the start of the calendar year on 1 Caitra, we ought to subtract from its longitude at mean Meṣasaṅkrānti an amount corresponding to its motion in 1 + 9/400 synodic months = 1227/40 *tithis*. This quantity converts (by the ratio of the number of *tithis* T in a *kalpa* to the corresponding number of civil days $C = T - U$) into 30; 11, 42, … days. This number of days multiplied by the planet's mean daily motion $\bar{v}$ (listed in column 2 of Schema 15) is the quantity to be subtracted from the mean Meṣasaṅkrānti longitude in order to produce the desired epoch mean longitude for 1 Caitra.

Unfortunately, the reconstructed calculations following this rationale produce results that agree only to the nearest integer degree with Bhāskara's listed offsets, as shown in column 3 of Schema 15. If the estimated time interval in days between 1 Caitra and Meṣasaṅkrānti is changed from 30; 11, 42 days to an empirically adjusted value of 30; 8, 47 days, however, most of the results agree with Bhāskara's to the nearest arcsecond (see column 4 of Schema 15). (Correcting the small longitudinal decrements for that time interval by the corresponding Brāhma-*bījas* would change the results by at most one arcsecond, so this was not done for any of the reconstructed values.) We do not know exactly what combination of differences in computational precision and in choice of parameters may best account for these small discrepancies.

2.2.2 Mean Longitude Increments

After specifying the planets' epoch mean longitudes, Bhāskara proceeds in *Karaṇakutūhala* 1.7–12 (Mishra, 1991, pp. 5–11) to state a set of approximate formulae for computing the longitudinal increments $\bar{v} \cdot c$, which are to be added to the epoch mean longitudes $\bar{\lambda}_E$ to produce the current mean longitudes of the planets at the end of the user's *ahargaṇa* c. With the partial exception of the formula for the Sun, all Bhāskara's $\bar{v} \cdot c$ formulae consist simply of approximating and modifying the value of $\bar{v}$ in degrees per day, both to make it easier to compute with than the cumbersome sexagesimal fractions shown in column 2 of Schema 15, and to take into account the planet's Brāhma-*bīja* correction

SCHEMA 14 Canonical data involved in planetary mean longitude computations in the *Karaṇakutūhala*. Column 2: Bhāskara's epoch mean longitudes $\bar{\lambda}_E$ for the planets as stated in *KKu* 1.4–6. Column 3: The Brāhmapakṣa's standard values for planetary revolutions K in a *kalpa* (*Brāhmasphuṭasiddhānta* 1.15–20 (Dvivedī, 1901–1902, pp. 5–6), *Siddhāntaśiromaṇi* 1.2.1–6 (Śāstrī, 1989, p. 10)). Column 4: The Brāhmapakṣa's annual *bīja* corrections applied to planetary mean longitudes during the Kaliyuga

Planet	$\bar{\lambda}_E$ (*KKu* 1.4–6)	K	Brāhma-*bīja* (′/year)
Sun	$10^s, 29°; 13, 0$	4,320,000,000	−3/200
Moon	$10^s, 29°; 5, 50$	57,753,300,000	−5/200
Moon's apogee	$4^s, 15°; 12, 59$	488,105,858	−2/200
Moon's node	$(-)9^s, 17°; 25, 9$	(−)232,311,168	(−)2/200
Mars	$7^s, 21°; 24, 21$	2,296,828,522	+1/200
Mercury's *śīghra*	$2^s, 21°; 14, 30$	17,936,998,984	+52/200
Jupiter	$2^s, 4°; 0, 51$	364,226,455	−5/200
Venus's *śīghra*	$8^s, 18°; 5, 55$	7,022,389,492	−15/200
Saturn	$4^s, 3°; 43, 17$	146,567,298	+4/200

SCHEMA 15 Reconstructed values of data pertaining to mean longitude computations in the *Karaṇakutūhala*. Column 2: Approximate daily mean motions for the planets computed from K and C. Column 3: Reconstructed planetary epoch mean longitudes $\bar{\lambda}_E$ assuming 30; 11, 42 days as the interval from 1 Caitra to Meṣasaṅkrānti Śaka 1105. Column 4: Reconstructed planetary epoch mean longitudes $\bar{\lambda}_E$ assuming 30; 8, 47 days as the interval from 1 Caitra to Meṣasaṅkrānti Śaka 1105

Planet	$\bar{v}$ (°/d)	$\bar{\lambda}_E$ (I)	$\bar{\lambda}_E$ (II)
Sun	0; 59, 8, 10, 21	$10^s, 29°; 10, 7, 15$	$10^s, 29°; 12, 59, 48$
Moon	13; 10, 34, 53, 0	$10^s, 28°; 27, 16, 51$	$10^s, 29°; 5, 43, 40$
Moon's apogee	0; 6, 40, 53, 56	$4^s, 15°; 12, 38, 45$	$4^s, 15°; 12, 58, 15$
Moon's node	(−)0; 3, 10, 48, 20	$(-)9^s, 17°; 24, 59, 44$	$(-)9^s, 17°; 25, 9, 1$
Mars	0; 31, 26, 28, 7	$7^s, 21°; 22, 50, 7$	$7^s, 21°; 24, 21, 52$
Mercury's *śīghra*	4; 5, 32, 18, 28	$2^s, 21°; 2, 31, 7$	$2^s, 21°; 14, 27, 34$
Jupiter	0; 4, 59, 9, 9	$2^s, 4°; 0, 37, 3$	$2^s, 4°; 0, 51, 36$
Venus's *śīghra*	1; 36, 7, 44, 35	$8^s, 18°; 1, 14, 24$	$8^s, 18°; 5, 54, 53$
Saturn	0; 2, 0, 22, 51	$4^s, 3°; 43, 10, 53$	$4^s, 3°; 43, 16, 44$

shown in column 4 of Schema 14. As reconstructed in the analysis laid out in the remainder of this section, the creation of the formulae involved the following procedures on Bhāskara's part:

- **Derive the first term(s) of the $\bar{\nu}$ approximation by a continued fraction**
  The *kalpa*-parameter expression 360 K/C is reduced to a continued fraction which is approximated by a simpler fraction, in units of degrees per day.
- **Compute the correction term for the simpler fraction approximation**
  The difference between the original exact expression $\bar{\nu} = 360\,K/C$ and the abovementioned term approximating $\bar{\nu}$ is calculated. This correction term has positive sign if the approximation is too small and negative sign if the approximation is too big.
- **Compute the corresponding value of the Brāhma-*bīja* correction**
  The annual Brāhma-*bīja* is converted to the appropriate units, usually degrees per day.
- **Combine the correction term and the Brāhma-*bīja* term**
  The fractions representing the first correction term and the Brāhma-*bīja* are arithmetically combined and modified to a simpler form, again by approximating a continued fraction.
- **If required, adjust the formula's fractional term(s) so they have the same numerator**
  For ease of verbal expression in Sanskrit verse, and/or for stylistic elegance and/or some other reason, Bhāskara generally prefers fractional terms in the same formula to share a common numerator, and apparently modifies some of them slightly in order to effect this.

2.2.3 Mean Longitude Increment for the Sun (and the Bodies of Mercury and Venus)

ahargaṇo viśvaguṇas trikhāṃkair
bhaktaḥ phalono dyugaṇo lavādyāḥ ‖
ravijñaśukrāḥ syur athābdavṛndād *upajāti*
vedāṃgalabdhena kalādinonāḥ ‖ 1.7 ‖ (*haṃsī*)

VERSE 1.7

The *ahargaṇa* is multiplied by 13 [and] divided by 903. The *ahargaṇa* diminished by the result is [the mean longitude increment of] the Sun, Mercury, and Venus, beginning with degrees. These should be diminished by the minutes etc. of a 64th part of the accumulated years.

Equivalent formula:

$$\bar{\lambda} = \bar{\lambda}_E + \left(c - \frac{13c}{903}\right)^{(\circ)} - \frac{y}{64}^{(\prime)}$$

First approximation term, in degrees per day:

$$\frac{360\,K}{C} = \frac{360 \cdot 4320000000}{1577916450000} = \frac{384000}{389609} = \cfrac{1}{1 + \cfrac{1}{68 + \cfrac{1}{2 + \cfrac{1}{5 + \cfrac{423}{433}}}}}$$

$$\approx \frac{890}{903} = 1 - \frac{13}{903}$$

Correction term for accuracy:[7]

$$360 - \frac{890}{903} \cdot 6, 5; 15, 30, 22, 30 \approx -0; 0, 0, 2, 15 = -\frac{9}{4} \cdot \frac{1'}{3600} = -\frac{1'}{1600}$$

Combination of correction term for accuracy with Brāhma-*bīja* correction, in arcminutes per year:

$$-\frac{1'}{1600} + \left(-\frac{3'}{200}\right) = -\frac{25'}{1600} = -\frac{1'}{64}$$

2.2.4 Mean Longitude Increment for the Moon

ahnāṃ gaṇaḥ śakraguṇo vihīnaḥ
svātyaṣṭi bhāgena lavādir induḥ ||
ahargaṇāt khābhrarasāṣṭabhaktād *upajāti*
āptena bhāgādi phalena hīnaḥ || 1.8 || (*sālā*)

7 S. Dvivedi (Mishra, 1991, pp. 7–8) explains the term $y/64$ (which does not take into account the current elapsed fraction of a year at the user's chosen date) by the following reasoning: The actual increment in mean solar longitude over a solar year is exactly 360°, whereas the continued-fraction approximation $\bar{\nu} \approx 890/903$ multiplied by the year-length in civil days $C/Y = 6, 5; 15, 30, 22, 30$ is approximately $6, 0; 0, 0, 2, 15$ degrees. In the formulae for all the other planets, the Brāhma-*bīja* term in arcminutes per year is converted into degrees per day by dividing it by $60\,C/Y = 3506481/160$.

VERSE 1.8

> The *ahargaṇa*, multiplied by 14 and decreased by its own 17th part, is [the mean longitude increment of] the Moon, beginning with degrees. [That] is decreased by the result in degrees etc. obtained from the *ahargaṇa* divided by 8600.

Equivalent formula:

$$\bar{\lambda} = \bar{\lambda}_E + \left(14c - \frac{14c}{17}\right) - \frac{c}{8600}$$

First approximation term, in degrees per day:

$$\frac{360\,K}{C} = \frac{360 \cdot 57753300000}{1577916450000} = \frac{15400880}{1168827} = 13 + \cfrac{1}{5 + \cfrac{1}{1 + \cfrac{1}{2 + \cfrac{2288}{67947}}}}$$

$$\approx 13 + \frac{3}{17} \quad = 14 - \frac{14}{17}$$

Correction term for accuracy:

$$\frac{15400880}{1168827} - \left(14 - \frac{14}{17}\right) = \frac{15400880}{1168827} - \frac{224}{17} = -\frac{208}{1806369}$$

Correction term for Brāhma-*bīja*:

$$-\frac{5}{200}\,({}'/\text{year}) = -\frac{5}{200 \cdot 60\,C/Y}\,({}^{\circ}/\text{day}) = -\frac{5}{200 \cdot \frac{3506481}{160}} = -\frac{4}{3506481}$$

Combination and simplification of the correction terms, in degrees per day:

$$-\frac{208}{1806369} + \left(-\frac{4}{3506481}\right) = -\frac{6932}{59610177} = -\frac{1}{8599 + \frac{1909}{6932}} \approx -\frac{1}{8600}$$

2.2.5 Mean Longitude Increment for the Moon's Apogee

gaṇo dvidhā gobhir inābhravedair *upajāti*
labdhaikyam aṃśādi bhaved vidhūccam ‖ 1.9ab ‖ (*haṃsī*)

VERSE 1.9ab

The *ahargaṇa* is put down twice; the sum of the quotients [of those two, divided respectively] by 9 [and] by 4012, beginning with degrees, is [the mean longitude increment of] the Moon's apogee.

Equivalent formula:

$$\bar{\lambda} = \bar{\lambda}_E + \frac{c}{9} + \frac{c}{4012}$$

First approximation term, in degrees per day:

$$\frac{360\,K}{C} = \frac{360 \cdot 488105858}{1577916450000} = \frac{244052929}{2191550625} = \frac{1}{8 + \frac{239127193}{244052929}} \approx \frac{1}{9}$$

Correction term for accuracy:

$$\frac{244052929}{2191550625} - \frac{1}{9} = \frac{547304}{2191550625}$$

Correction term for Brāhma-*bīja*:

$$-\frac{2}{200}\,({}'/\text{year}) = -\frac{2}{200 \cdot 60\,C/Y}\,({}^{\circ}/\text{day}) = -\frac{2}{200 \cdot \frac{3506481}{160}} = -\frac{8}{17532405}$$

Combination and simplification of the correction terms, in degrees per day:

$$\frac{547304}{2191550625} + \left(-\frac{8}{17532405}\right) = \frac{49664}{199231875} = \frac{1}{4011 + \frac{29571}{49664}} \approx \frac{1}{4012}$$

2.2.6 Mean Longitude Increment for the Moon's Node

dvidhāṅkacandraiḥ khakhabhair dinaughād *upajāti*
āptāṃśayogo bhavatīndupātaḥ ‖ 1.9cd ‖ (*haṃsī*)

VERSE 1.9cd

The sum of the degrees from dividing the *ahargaṇa* put down twice by 19 [and] by 2700 [respectively] is [the mean longitude increment of] the Moon's node.

Equivalent formula:
(We follow Bhāskara's example as described in note 5 by treating all these terms as positive, disregarding the node's backwards rotation.)

$$\bar{\lambda} = \bar{\lambda}_E + \frac{c}{19} + \frac{c}{2700}$$

First approximation term, in degrees per day:

$$\frac{360\,K}{C} = \frac{360 \cdot 232311168}{1577916450000} = \frac{116155584}{2191550625} = \frac{1}{18 + \dfrac{100750113}{116155584}} \approx \frac{1}{19}$$

Correction term for accuracy:

$$\frac{116155584}{2191550625} - \frac{1}{19} = \frac{1711719}{46266068 75}$$

Correction term for Brāhma-*bīja*:

$$\frac{2}{200}\,({}'/\text{year}) = \frac{2}{200 \cdot 60\,C/Y}\,(\circ/\text{day}) = \frac{2}{200 \cdot \dfrac{3506481}{160}} = \frac{8}{17532405}$$

Combination and simplification of the correction terms, in degrees per day:

$$\frac{1711719}{46266068 75} + \frac{8}{17532405} = \frac{15424471}{41639461875} = \frac{1}{2699 + \dfrac{8814646}{15424471}} \approx \frac{1}{2700}$$

2.2.7 Mean Longitude Increment for Mars

rudraghno dyucayo dvidhā śaśiyamair
vedābdhisiddheṣubhir
bhakto 'ṃśādi phaladvayaṃ tu sahitaṃ syān
medinīnandanaḥ ‖ 1.10ab ‖ *śārdūlavikrīḍita*

VERSE 1.10ab

The *ahargaṇa* multiplied by 11 is put down twice [and] divided [separately] by 21 [and] by 52444. The two results in degrees etc., added together, are [the mean longitude increment of] Mars.

Equivalent formula:

$$\bar{\lambda} = \bar{\lambda}_E + \frac{11C}{21} + \frac{11C}{52444}$$

First approximation term, in degrees per day:

$$\frac{360\,K}{C} = \frac{360 \cdot 2296828522}{1577916450000} = \frac{1148414261}{2191550625} = \cfrac{1}{1 + \cfrac{1}{1 + \cfrac{1}{9 + \cfrac{1}{1 + \cfrac{9642606}{95635291}}}}} \approx \frac{11}{21}$$

Correction term for accuracy:

$$\frac{1148414261}{2191550625} - \frac{11}{21} = \frac{3214202}{15340854375}$$

Correction term for Brāhma-*bīja*:

$$\frac{1}{200}\,(\prime/\text{year}) = \frac{1}{200 \cdot 60\,C/Y}\,(^\circ/\text{day}) = \frac{1}{200 \cdot \frac{3506481}{160}} = \frac{4}{17532405}$$

Combination and simplification of the correction terms, in degrees per day:[8]

$$\frac{3214202}{15340854375} + \frac{4}{17532405} = \frac{3217702}{15340854375}$$

$$= 11 \cdot \frac{3217702}{168749398125}$$

$$= 11 \cdot \frac{1}{52444 + \dfrac{234437}{3217702}} \approx \frac{11}{52444}$$

2.2.8 Mean Longitude Increment for Mercury's *śīghra*

vedaghno dyucayaḥ svakīya dahīnābdhy
aṃśena yukto bhaved
bhāgādikṣvacalaṃ gaṇāt
kṣitiyamendrāptāṃśakair varjitam ‖ 1.10cd ‖ *śārdūlavikrīḍita*

VERSE 1.10cd

The *ahargaṇa* multiplied by 4, added to its own 43rd part, in degrees etc., is [the mean longitude increment of] the *śīghra* of Mercury. [It] is diminished by the degrees of the *ahargaṇa* divided by 1421.

Equivalent formula:

$$\bar{\lambda} = \bar{\lambda}_E + \left(4c + \frac{4c}{43}\right) - \frac{c}{1421}$$

First approximation term, in degrees per day:[9]

8 This appears to be one of the instances where Bhāskara modified the computation of his second continued-fraction approximation to make the numerator of the resulting fraction equal to that of the first term. The nearest continued-fraction approximation to the original sum of the correction terms would have yielded a fraction with a different numerator:

$$\frac{3217702}{15340854375} = \cfrac{1}{4767 + \cfrac{1}{1 + \cfrac{1}{1 + \cfrac{1}{1 + \cfrac{1}{4 + \cfrac{5856}{228581}}}}}} \approx \frac{14}{66747}$$

9 Here again, Bhāskara seems to have tweaked the more accurate fractional part 6/65 of this term to the nearest equivalent with numerator equal to the term's integer part 4.

$$\frac{360\,K}{C} = \frac{360 \cdot 17936998984}{1577916450000} = \frac{8968499492}{2191550625} = 4 + \cfrac{1}{10 + \cfrac{1}{1 + \cfrac{1}{4 + \cfrac{33715557}{33716287}}}}$$

$$\approx 4 + \frac{6}{65} \quad \approx 4 + \frac{4}{43}$$

Correction term for accuracy:

$$\frac{8968499492}{2191550625} - \left(4 + \frac{4}{43}\right) = \frac{8968499492}{2191550625} - \frac{176}{43} = -\frac{67431844}{94236676875}$$

Correction term for Brāhma-*bīja*:

$$\frac{52}{200}\,({}'/\text{year}) = \frac{52}{200 \cdot 60\,C/Y}\,({}^\circ/\text{day}) = \frac{52}{200 \cdot \frac{3506481}{160}} = \frac{208}{17532405}$$

Combination and simplification of the correction terms, in degrees per day:

$$-\frac{67431844}{94236676875} + \frac{208}{17532405} = -\frac{66313844}{94236676875} = -\frac{1}{1421 + \frac{4704551}{66313844}} \approx -\frac{1}{1421}$$

2.2.9 Mean Longitude Increment for Jupiter

gaṇo dvidhārkair bhayamābdhibhiś ca *upajāti*
bhaktaḥ phalāṃśāntaram indramantrī ‖ 1.11ab ‖ (*haṃsī*)

VERSE 1.11ab

The *ahargaṇa* put down twice is divided [separately] by 12 and by 4227; the difference of the degrees of the results is [the mean longitude increment of] Jupiter.

Equivalent formula:

$$\bar{\lambda} = \bar{\lambda}_E + \frac{c}{12} - \frac{c}{4227}$$

First approximation term, in degrees per day:

$$\frac{360\,K}{C} = \frac{360 \cdot 364226455}{1577916450000} = \frac{72845291}{876620250} = \frac{1}{12 + \frac{2476758}{72845291}} \approx \frac{1}{12}$$

Correction term for accuracy:

$$\frac{72845291}{876620250} - \frac{1}{12} = -\frac{412793}{1753240500}$$

Correction term for Brāhma-*bīja*:

$$-\frac{5}{200}^{(\prime/\text{year})} = -\frac{5}{200 \cdot 60\,C/Y}^{(\circ/\text{day})} = -\frac{5}{200 \cdot \frac{3506481}{160}} = -\frac{4}{3506481}$$

Combination and simplification of the correction terms, in degrees per day:

$$-\frac{412793}{1753240500} + \left(-\frac{4}{3506481}\right) = -\frac{414793}{1753240500} = -\frac{1}{4226 + \frac{325282}{414793}} \approx -\frac{1}{4227}$$

2.2.10 Mean Longitude Increment for Venus's *śīghra*

nṛpāhato 'hnāṃ nicayo dvidhāsau — *upajāti*
bhūbāṇavedādribhir abhracandraiḥ ‖ 1.11cd ‖ — (*haṃsī*)

bhakto lavādyaṃ phalayor yadaikyaṃ
tajjāyate daityaguroś caloccam ‖ 1.12ab ‖ — *indravajrā*

VERSES 1.11cd–12ab

The *ahargaṇa* is multiplied by 16 and put down twice. That is divided [separately] by 7451 [and] by 10. Whatever is the sum of the two results, beginning with degrees, that becomes [the mean longitude increment of] the *śīghra* of Venus.

Equivalent formula:

$$\bar{\lambda} = \bar{\lambda}_E + \frac{16c}{10} + \frac{16c}{7451}$$

First approximation term, in degrees per day:

$$\frac{360\,K}{C} = \frac{360 \cdot 7022389492}{1577916450000} = \frac{3511194746}{2191550625} = 1 + \cfrac{1}{1 + \cfrac{1}{1 + \cfrac{1}{1 + \cfrac{424168887}{447737617}}}}$$

$$\approx \frac{8}{5} = \frac{16}{10}$$

Correction term for accuracy:

$$\frac{3511194746}{2191550625} - \frac{16}{10} = \frac{4713746}{2191550625}$$

Correction term for Brāhma-*bīja*:

$$-\frac{15}{200}\,(\prime/\text{year}) = -\frac{15}{200 \cdot 60\,C/Y}\,(\circ/\text{day}) = -\frac{15}{200 \cdot \frac{3506481}{160}} = -\frac{4}{1168827}$$

Combination and simplification of the correction terms, in degrees per day:[10]

10 Here is another instance where Bhāskara appears to have modified his fractional terms to give them the same numerator. The first approximation term could be expressed as either 16/10 or 8/5, but the approximate combined correction term $\frac{3}{1397} = \frac{48}{22352}$ cannot be exactly converted to any fraction with either 16 or 8 in the numerator. Its nearest approximations satisfying that condition are

$$\frac{48}{22352} \approx \frac{48}{22353} = \frac{16}{7451} \quad \text{and} \quad \frac{48}{22352} \approx \frac{48}{22350} = \frac{8}{3725},$$

of which the former is slightly more accurate.

$$\frac{4713746}{2191550625} + \left(-\frac{4}{1168827}\right) = \frac{4706246}{2191550625} = \cfrac{1}{465 + \cfrac{1}{1 + \cfrac{1}{2 + \cfrac{26213}{1560011}}}}$$

$$\approx \frac{3}{1397} = \frac{48}{22352} \approx \frac{48}{22353} = \frac{16}{7451}$$

2.2.11 Mean Longitude Increment for Saturn

bhakto 'bhrāmais turaṅgāṃgarāma-
nandair dvidhāṃśādi phalaikyam ārkiḥ ‖ 1.12cd ‖ *indravajrā*

Verse 1.12cd

[The *ahargaṇa*] put down twice is divided by 30 [and] by 9367. The sum of the results in degrees etc. is [the mean longitude increment of] Saturn.

Equivalent formula:

$$\bar{\lambda} = \bar{\lambda}_E + \frac{c}{30} + \frac{c}{9367}$$

First approximation term, in degrees per day:

$$\frac{360\,K}{C} = \frac{360 \cdot 146567298}{1577916450000} = \frac{73283649}{2191550625} = \cfrac{1}{1 + \cfrac{1}{29 + \cfrac{66324804}{73283649}}} \approx \frac{1}{30}$$

Correction term for accuracy:

$$\frac{73283649}{2191550625} - \frac{1}{30} = \frac{51547}{487011250}$$

Correction term for Brāhma-*bīja*:

$$\frac{4}{200}\,({}'/\text{year}) = \frac{4}{200 \cdot 60\,C/Y}\,({}^\circ/\text{day}) = \frac{4}{200 \cdot \frac{3506481}{160}} = \frac{16}{17532405}$$

Combination and simplification of the correction terms, in degrees per day:

$$\frac{51547}{487011250} + \frac{16}{17532405} = \frac{467923}{4383101250} = \frac{1}{9367 + \dfrac{66509}{467923}} \approx \frac{1}{9367}$$

2.3 *The Longitudinal-Difference Correction or* deśāntara

As noted in *Karaṇakutūhala* 1.6 discussed above, the planetary mean longitudes by default are computed for sunrise at the notional Indian zero-point of terrestrial latitude and longitude, i.e., the ideal position of Laṅkā at the intersection of the earth's equator and its prime meridian passing through Ujjayinī, modern Ujjain; *vide* (Rai, 1974). Bhāskara's next two verses (not quoted here) recapitulate key data for finding the corresponding planetary mean longitudes for an observer at a different terrestrial longitude. First, verse 1.13 states standard values of planetary mean daily motion $\bar{\nu}$ in arcminutes and arcseconds: they agree to the nearest arcsecond with the recomputed values in column 2 of Schema 15, except for the mean daily motion of Jupiter which is stated as 5 rather than 4; 59 arcminutes. Verse 1.14 then lists various cities considered in traditional Indian geography to lie on the prime meridian between Laṅkā and Mount Meru at the north pole, presumably to assist the user in determining the distance between their own location and the prime meridian. This information is employed in computing the so-called *deśāntara* ('locality difference') or longitudinal-difference correction, defined in *Karaṇakutūhala* 1.15 (Mishra, 1991, p. 12) as follows:

> *rekhā svadeśāntarayojanaghnī*
> *gatir grahasyābhragajair vibhaktā* ||
> *labdhā viliptā khacare vidheyā* — *upajāti*
> *prācyām ṛṇaṃ paścimato dhanaṃ tāḥ* || 1.15 || — (*vāṇī*)

VERSE 1.15

> The velocity of the planet is multiplied by the distance in *yojanas* of one's own locality from the prime meridian, and divided by 80. The resulting arcseconds are applied to [the longitude of] the planet, negatively if east, positively if west.

That is, the *deśāntara* correction in arcseconds is given by

$$deśāntara^{('')} = \frac{\bar{\nu} \cdot d}{80} \quad \text{or} \quad \frac{deśāntara^{('')}}{\bar{\nu}} = \frac{d}{80},$$

where d is the difference in terrestrial longitude between the user's locality and the prime meridian, in units of *yojanas* (a linear measure on the order of 10 kilometers; *vide* (González-Reimann, 2009, p. 424)). We can explain this relation in terms of the spherical geometry of the earth as follows: The quantity that results from dividing the *deśāntara* in arcseconds by $\bar{\nu}$ in arcminutes per day is a time difference in units of sixtieths of a day or *ghaṭikās*. If that time difference were one entire daily revolution of the celestial equator, i.e., 60 *ghaṭikās*, then the proportion would become

$$60 = \frac{d}{80} \quad \text{or} \quad d = 4800,$$

implying that the complete circumference of the earth traversed in one daily revolution is 4800 *yojanas*.

The earth's equatorial circumference is generally assumed in Indian astronomy to be about 5000 *yojanas* (Misra, 2016, pp. 126–127). Bhāskara states in *Siddhāntaśiromaṇi* 1.7.1 (Śāstrī, 1989, p. 36) that it is 4967 *yojanas*, in which case 4800 *yojanas* would represent the length of the parallel of latitude at almost 15°N. For an observer located at that terrestrial latitude and d *yojanas* away from the prime meridian in longitude, a planet moving with mean velocity $\bar{\nu}$ arcminutes per day will traverse $\bar{\nu} \cdot d/4800$ arcminutes, or equivalently $\bar{\nu} \cdot d/80$ arcseconds, of celestial mean longitude; for observers on a smaller or larger parallel of latitude, of course, the result will be only approximate. This *deśāntara* is applied to the planet's mean longitude positively or negatively when the observer's locality is west or east, respectively, of the prime meridian.

2.4 *The Annual Correction or* abdabīja

Bhāskara further elaborates the ingenious formulae for planetary mean longitude increments discussed above with some additional (optional?) smaller correction terms for all the planets except the Sun, Mars and Saturn. Amounting only to a fraction of an arcsecond for every year elapsed since the epoch, these *abdabījas* ('annual *bījas*') are specified in *Karaṇakutūhala* 1.16 (Mishra, 1991, p. 13) and summarized in Schema 16:

abdā gajāśvaistrirasair vibhājitā
ṛṇaṃ viliptāsu śaśījyayoḥ kramāt ||

viśvaiḥ kharāmair dviyamaiś ca khecaraiḥ
pātoccasaumyāsphujitāṃ dhanaṃ tathā ‖ 1.16 ‖ *indravaṃśā*

VERSE 1.16

The years divided by 78 [and] by 63 are [applied] negatively to the arcseconds of [the longitude of] the Moon and of Jupiter, respectively; divided by 13, by 30, by 22, and by 9, they are [applied] positively to [the longitudes of] the node, the apogee, [the *śīghra* of] Mercury, and [the *śīghra* of] Venus, in the same way.

SCHEMA 16 Additional annual correction factors (*abdabīja*) stated in *KKu* 1.16

Planet	*abdabīja* factor (″/year)
Sun	(none)
Moon	−1/78
Moon's apogee	+1/30
Moon's node	(−)1/13
Mars	(none)
Mercury's *śīghra*	+1/22
Jupiter	−1/63
Venus's *śīghra*	+1/9
Saturn	(none)

These *abdabījas* can be explained mathematically as refinements to the continued-fraction approximations of the combined correction terms in the mean longitudinal increment formulae. The difference between the exact and approximate values of a planet's combined correction term is converted from degrees per day to arcseconds per year by multiplying it by $3600\,C/Y = 10519443/8$. Then this finer correction term is itself approximated by rounding off a continued fraction. The following equations briefly sketch the reconstruction of Bhāskara's *abdabīja* corrections; it is not obvious to us why he considered this additional modification negligible in the case of the Sun, Mars and Saturn.

2.4.1 The *abdabīja* Factor for the Moon, in Arcseconds per Year

$$-\frac{1}{8599+\frac{1909}{6932}}-\left(-\frac{1}{8600}\right)=-\frac{5023}{512647522200}\ (^\circ/\text{day})$$

$$=-\frac{15069}{1169600}\ (''/\text{year})=-\frac{1}{77+\frac{9287}{15069}}\approx-\frac{1}{78}$$

2.4.2 The *abdabīja* Factor for the Moon's Apogee, in Arcseconds per Year

$$\frac{1}{4011+\frac{29571}{49664}}-\frac{1}{4012}=\frac{20093}{799318282500}\ (^\circ/\text{day})$$

$$=\frac{663069}{20060000}\ (''/\text{year})=\frac{1}{30+\frac{167930}{663069}}\approx\frac{1}{30}$$

2.4.3 The *abdabīja* Factor for the Moon's Node, in Arcseconds per Year (Again, we follow Bhāskara in treating this quantity as positive for computational purposes.)

$$\frac{1}{2699+\frac{8814646}{15424471}}-\frac{1}{2700}=\frac{29377}{499673542500}\ (^\circ/\text{day})$$

$$=\frac{29377}{380000}\ (''/\text{year})=\frac{1}{12+\frac{27476}{29377}}\approx\frac{1}{13}$$

2.4.4 The *abdabīja* Factor for Mercury's *śīghra*, in Arcseconds per Year

$$-\frac{1}{1421+\frac{4704551}{66313844}}-\left(-\frac{1}{1421}\right)=\frac{4704551}{133910317839375}\ (^\circ/\text{day})$$

$$=\frac{14113653}{305515000}\ (''/\text{year})=\frac{1}{21+\frac{9128287}{14113653}}\approx\frac{1}{22}$$

2.4.5 The *abdabīja* Factor for Jupiter, in Arcseconds per Year

$$-\frac{1}{4226+\frac{325282}{414793}}-\left(-\frac{1}{4227}\right)=-\frac{29837}{2470315864500}\ (\circ/\text{day})$$

$$=-\frac{89511}{5636000}\ (''/\text{year})$$

$$=-\frac{1}{62+\frac{86318}{89511}}\approx-\frac{1}{63}$$

2.4.6 The *abdabīja* Factor for Venus's *śīghra*, in Arcseconds per Year

$$\frac{4706246}{2191550625}-\frac{16}{7451}=\frac{1428946}{16329243706875}\ (\circ/\text{day})$$

$$=\frac{2143419}{18627500}\ (''/\text{year})=\frac{1}{8+\frac{1480148}{2143419}}\approx\frac{1}{9}$$

With the *abdabīja* correction the planetary mean longitude computation in the *Karaṇakutūhala* is considered complete. The remaining modifications to these quantities are dealt with in the handbook's chapter 2 on true longitudes, discussed in the following sections.

2.5 *The Basic Models and Parameters of Planetary Orbital Inequalities*

In order to determine the true longitudes of the planets, their mean longitudes need to be adjusted for the inequalities of their orbits. These inequalities are represented in Sanskrit geocentric astronomy by models qualitatively similar to those of its Ptolemaic counterpart in their use of non-concentric circles (generally eccentrics or epicycles), each possessing an apogee (*ucca*) or point on the circle farthest from the earth.

The first of these corrections is the so-called *manda* ('slow') one, which corresponds to the assumption, in an eccentric geocentric orbital model such as the one illustrated in figure 15 in Appendix, section 2.7, that the orbiting body is moving with uniform velocity upon a circle whose center is displaced from the earth by an amount of eccentricity r_M. This displacement produces changes in speed and position over the course of the body's revolution that are qualitatively similar to the effect of an elliptical orbit with the earth at one focus. The '*manda*-anomaly', here denoted κ_M, is the difference between the mean position of the planet (which in the case of Mercury or Venus, as previously noted, is that of the mean Sun) and that of its *manda*-apogee. The *manda*-anomaly and the eccentricity r_M determine the *manda*-equation or correction μ (*mandaphala*), the required adjustment to the planet's mean longitude.

The second or *śīghra* ('fast') inequality models the synodic periods of the star-planets. Namely, it accounts for the phenomena of planetary stations and retrogradation, heliocentrically explained by the fact that the other planets (in the modern sense) as well as the earth are revolving about the Sun. Thus a planet seen from the earth as they pass in their orbits can appear to pause and go backwards temporarily, as illustrated in figure 16 in Appendix, section 2.8. Analogously to the *manda*, it is the *śīghra*-anomaly κ_S or longitudinal interval between the *śīghra*-apogee and the (*manda*-corrected) planetary longitude, along with the *śīghra*-circle radius r_S, that determines the amount of the *śīghra*-equation or correction σ (*śīghraphala*). The Sun and Moon, not possessing what we nowadays term a heliocentric orbit, are affected only by the *manda*-inequality and thus are not *śīghra*-corrected.

The first requirement for computing true-longitude corrections according to these models is to state the parameters for the orbital inequalities (reproduced in Schema 17) and the calculation of their respective anomalies. This is what Bhāskara does in *Karaṇakutūhala* 2.1–4 (Mishra, 1991, p. 39), beginning as follows:

mandoccam arkasya gajādribhāgā
bhaumādikānāṃ sadalāṣṭasūryāḥ ||
tattvāśvinaḥ sārddhayamādricandrāḥ
kvaṣṭau śaśāṅkāṅgayamāḥ krameṇa || 2.1 || *indravajrā*

VERSE 2.1

[The longitude of] the Sun's *manda*-apogee is 78 degrees; [those] of Mars etc. are 128 $\frac{1}{2}$, 225, 172 $\frac{1}{2}$, 81, [and] 261 respectively.

This first verse lists the (effectively unchanging) longitudes of the *manda*-apogees of the planets other than the Moon (*vide* Schema 17), which are considered to move with extremely slow constant angular velocity during a *kalpa*; they are not substantially different from the values given in Brāhmapakṣa texts centuries earlier.[11] The *manda*-eccentricity values r_M, on the other hand, are

11 The Brāhmapakṣa's *manda*-apogee longitudes are discussed in (Pingree, 1978, pp. 556, 568); Bhāskara in *Karaṇakutūhala* 2.1 merely rounds these values to the nearest half-degree.

SCHEMA 17 *Manda*-apogee longitudes (for planets other than the Moon) and *śīghra*-radii (for the five star-planets) from *Karaṇakutūhala* 2.1–2

Planet	*manda*-apogee longitude (*KKu* 2.1)			*śīghra*-radius (*KKu* 2.2)
Sun	2^s	18°		
Mars	4^s	8°	30′	81
Mercury	7^s	15°		44
Jupiter	5^s	22°	30′	23
Venus	2^s	21°		87
Saturn	8^s	21°		13

nowhere stated in the *Karaṇakutūhala*. But as discussed in Appendix, section 2.7, their sizes are implied in the *Karaṇakutūhala*'s formulae for the *manda*-correction, and can be reconstructed from the equivalent parameters specified as circumferences of *manda*-epicycles in *Siddhāntaśiromaṇi* 2.22 (Śāstrī, 1989, p. 45); *vide* Schema 19.

The star-planets' *śīghra*-apogees, unlike the *manda*-apogees, must change position quickly to account for the planets' synodic phenomena. Their longitudes in the case of the inferior planets Mercury and Venus have been dealt with in the section on mean longitudes, while the mean Sun itself serves as the *śīghra*-apogee of the superior planets Mars, Jupiter and Saturn, as Bhāskara reminds us in *Karaṇakutūhala* 2.2. This verse also specifies the *paras* or radii r_S of the *śīghra*-circles (scaled to the trigonometric radius R, set at 120 in the *Karaṇakutūhala*; *vide* Schema 17), whose effect on the *śīghra*-correction is explained in Appendix, section 2.8.

kukuñjarā vedakṛtās tridasrāḥ
saptāhayo viśvamitāḥ parākhyāḥ ||
bhaumādikānām atha madhyamo 'rkaḥ *upajāti*
śīghroccam ijyāraśanaiś carāṇām || 2.2 || (*kīrti*)

VERSE 2.2

81, 44, 23, 87, 13 are called the '*paras*' [*śīghra*-circle radii] of Mars etc. Now the mean Sun is the *śīghra*-apogee of the planets [known] as Jupiter, Mars, and Saturn.

The computation of the orbital anomaly from the mean longitudes of planet and apogee, and its relation to the sign of the corresponding longitude correction, are then specified:

grahonam uccaṃ mṛdu cañcalaṃ ca
kendre bhavetāṃ mṛducañcalākhye ‖
tribhis tribhir bhaiḥ padam atra kalpyaṃ *upajāti*
svarṇaṃ phalaṃ meṣatulādikendre ‖ 2.3 ‖ (*sālā*)

Verse 2.3

[The longitude of] the *manda* or *śīghra* apogee [is] diminished by [that of] the planet; [the results] become the two anomalies called '*manda*' and '*śīghra*.' A quadrant [of anomaly] is here to be determined by [zodiacal] signs [taken] three by three. The equation is [to be] added or subtracted when the anomaly is [in the half-circles] beginning with Aries or Libra [respectively].

tryūnaṃ bhujaḥ syāt tryadhikena hīnaṃ
bhārddhaṃ ca bhārddhād adhikaṃ vibhārddhaṃ ‖
navādhikenonitam arkabhaṃ ca *upajāti*
bhavec ca koṭis trigṛhaṃ bhujonaṃ ‖ 2.4 ‖ (*rāmā*)

Verse 2.4

An arc less than three [signs] is [the desired arc]. Half a circle is diminished by [an arc] greater than three [signs], and [an arc] greater than half a circle is diminished by half a circle, and twelve signs are diminished by [an arc] greater than nine [signs, to give the desired arc]. And the complement should be three signs diminished by the arc.

That is, if the *manda* and *śīghra* inequalities are denoted by subscripts M and S and their apogees' longitudes by λ_{A_M} and λ_{A_S} respectively, their corresponding anomalies $\varkappa_M$ and $\varkappa_S$ should be found from

$$\varkappa_M = \lambda_{A_M} - \bar{\lambda},$$
$$\varkappa_S = \lambda_{A_S} - \lambda_M,$$

where $\bar{\lambda}$ as before is the planet's mean longitude and λ_M is its mean longitude after the *manda*-correction is applied.[12]

When the mean planet is less than 180° east of the apogee, its anomaly as computed by the above rule will be greater than 180°, so its equation will be in the western direction, i.e., negative; when the planet is less than 180° west of the apogee, the results are reversed (*vide* figure 15 in Appendix, section 2.7). 'Beginning with Aries' and 'beginning with Libra' are standard designations for any semicircular arcs starting at 0° and 180° respectively, even if they are not measured from the zero-point of the ecliptic. (Similarly, any semicircles extending from 90° to 270° and from 270° to 90° may be described respectively as 'beginning with Cancer' and 'beginning with Capricorn.') The reduction of arbitrary arcs to the first quadrant is self-explanatory.

2.6 *Using the* śīghra-*Anomaly of Mars to Correct Its* manda-*Apogee*

The *Karaṇakutūhala* procedure requires some final adjustments before embarking on the actual computation of the correction terms. Namely, in *Karaṇakutūhala* 2.5 (Mishra, 1991, p. 20) the longitude of the *manda*-apogee of Mars is presumed to be affected by its *śīghra*-anomaly:

bhaumāśu kendre padayātagamya
svalpasya liptāḥ khakhavedabhaktāḥ ||
labdhāṃśakaiḥ karkamṛgādi kendre
hīnānvitaṃ spaṣṭam asṛgmṛduccam || 2.5 || *indravajrā*

labdhāṃśakānāṃ trilavena hīnaḥ
spaṣṭaḥ paraḥ syāt kṣitinandanasya || 2.6ab || *indravajrā*

VERSES 2.5–6ab

In the case of Mars's *śīghra*-anomaly, the minutes of [whichever is] the lesser of [the part] past and [the part] to come of [its] quadrant are divided by 400. Depending on whether the anomaly is in [the semicircle] beginning with Cancer or Capricorn, [the longitude of] the *manda*-

12 Most earlier Sanskrit astronomy works, including Bhāskara's own *Siddhāntaśiromaṇi* in verse 2.18 (Śāstrī, 1989, p. 44), compute the two anomalies in opposite order: that is, they subtract λ_{A_M} from $\bar{\lambda}$ to obtain κ_M, but λ_M from λ_{A_S} to give κ_S. The *Karaṇakutūhala*'s more elegantly symmetrical algorithm allows Bhāskara to explain the determination of the anomaly and the sign of the equation in a single verse equally applicable to both inequalities.

> apogee of Mars diminished or increased [respectively] by the degrees of the quotient is correct. The *śīghra*-radius of Mars, diminished by a third part of the degrees of the quotient, should be correct.

The first of these algorithms amounts to multiplying the arc between the *śīghra*-anomaly of Mars and the closest integer multiple of 90° by the scale factor 3/20 or 9 arcminutes per degree. The result is applied to displace the longitude of Mars's *manda*-apogee backwards or forwards in the ecliptic, depending on whether the planet is in the half-circle of anomaly centered on opposition or in the one centered on conjunction, respectively. At its conjunction, opposition or quadrature there is no correction to the *manda*-apogee, while the correction is maximum ($0°;9 \times 45 = 6°;45$) when the anomaly is an odd multiple of 45°.

Qualitatively, this adjustment has the effect of moving the *manda*-apogee towards the planet at the octants around conjunction, which decreases the speed of its motion, and away from the planet at the octants around opposition, which increases its speed (or strictly speaking slows down its retrograde motion).[13]

In the next verse, one-third of the correction to the *manda*-apogee found by the previous rule is to be subtracted from the *śīghra*-radius of Mars specified in verse 2.2. As with the shifting of Mars's *manda*-apogee, rules for this sort of 'pulsation' of the size of epicycles go back to much earlier texts, although their theoretical significance is not entirely clear; cf. (Yano, 1997).

2.7 *The* manda-*Correction to Mean Longitude and Velocity*

After the foregoing preliminary modifications to the orbits, the *mandaphala* or *manda*-equation μ itself can finally be calculated. The diagram on the left in figure 15 shows μ as the angle $\angle \bar{P}\, O\, P$ between the planet's mean position(s) $\bar{P}$ with longitude $\bar{\lambda}$ and its corresponding corrected position P as viewed from the earth. The *manda*-corrected mean longitude λ_M of P will therefore be given by

$$\lambda_M = \bar{\lambda} \pm \mu.$$

13 Bhāskara evidently derived the rule from an algorithm in *Siddhāntaśiromaṇi* 2.24–25 (Śāstrī, 1989, pp. 46–47) in which the scale factor 6; 40/R sin 45 is applied to the R sine of the past or future arc of the quadrant of *śīghra*-anomaly (i.e., its sine scaled to the trigonometric radius R: *vide* Appendix, section 2.7). This normalizes the absolute value of the *manda*-apogee correction term to a maximum of 6°;40 at the octants of *śīghra*-anomaly, rather than 6°;45 as in the *Karaṇakutūhala*/*Brahmatulyasāraṇī* version. Bhāskara's commentary in the *Siddhāntaśiromaṇi* says of this correction merely *atrāgama eva pramāṇam*, i.e., it is the quantity prescribed in the received tradition. Its origin and efficacy are discussed in more detail in (Duke, 2005).

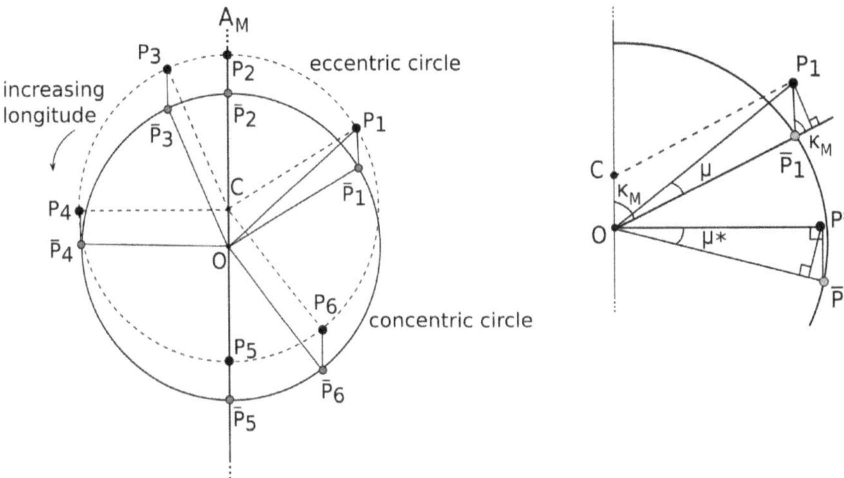

FIGURE 15 The *manda*-correction to mean longitude interpreted geometrically via an eccentric orbit. **Left:** The point O is the observer's position at the center of the concentric circle with radius R, on which the mean planet $\bar{P}$ moves eastward from position $\bar{P}_1$ to $\bar{P}_6$. The point C is the center of the eccentric circle with the same radius R, representing the actual path of the planet P moving from P_1 to P_6. The distance OC is the amount of eccentricity r_M, and A_M is the position of the *manda*-apogee from which the anomaly $\angle A_M O\bar{P}$ or $\varkappa_M$ is computed. **Right:** The *manda*-equation μ is derived trigonometrically from the right triangle with hypotenuse OP_1. Beneath it, the maximum equation $\mu*$ with R sine equal to $r_M = OC$ separates the planet's mean position $\bar{P}^*$ from its true position P^* when the anomaly is $\mu*$ degrees greater than 90°.

The angle μ depends trigonometrically on the size of the eccentricity $r_M = OC - \bar{P}P$ and the amount of the *manda*-anomaly $\varkappa_M = \angle A_M CP = \angle A_M O\bar{P}$. For instance, μ is zero whenever the planet is on the apsidal line defined by its apogee-point and the earth: i.e., when $\varkappa_M$ is 0 or 180 degrees. As Bhāskara explained in verse 2.4, the value of μ is positive (meaning that the planet's corrected longitude will be larger than its mean longitude) when the anomaly falls in the interval 0–180°, and negative thereafter.

The diagram on the right in figure 15 illustrates the trigonometric definition of the *mandaphala* μ as expressed in, e.g., *Siddhāntaśiromaṇi* 2.26–29 (Śāstrī, 1989, pp. 47–50), employing the so-called R sine and R cosine functions equivalent to the modern sine and cosine scaled to Bhāskara's chosen trigonometric radius $R = 120$. In the right triangle containing acute angle μ, its opposite side $R \sin \varkappa_M \cdot r_M / R$, and its adjacent side $R \pm R \cos \varkappa_M \cdot r_M / R$, the ratio of the opposite side to the hypotenuse is

SCHEMA 18 The *R* sine-difference values stated in *Karaṇakutūhala* 2.6cd–7 for each 10° of arc in the first quadrant, and their corresponding cumulative *R* sine values

Arc$^{(\circ)}$	*R* sine-difference	*R* sine
10	21	21
20	20	41
30	19	60
40	17	77
50	15	92
60	12	104
70	9	113
80	5	118
90	2	120

$$\sin\mu = \frac{R\sin\varkappa_M \cdot \frac{r_M}{R}}{\sqrt{\left(R\sin\varkappa_M \cdot \frac{r_M}{R}\right)^2 + \left(R \pm R\cos\varkappa_M \cdot \frac{r_M}{R}\right)^2}}.$$

The other right triangle beneath it represents the occurrence of the maximum absolute value $\mu*$ of the *manda*-equation when the *manda*-anomaly $\varkappa_M$ exceeds 90° by an amount equal to the arcsine of the eccentricity r_M, which is just $\mu*$ itself. By symmetry, the same maximum value of μ will also occur when $\varkappa_M$ falls short of 270° by that same amount $\mu*$:

$$\varkappa_M = \quad 90^\circ + \arcsin\left(\frac{r_M}{R}\right) \quad \text{or} \quad 270^\circ - \arcsin\left(\frac{r_M}{R}\right).$$

This trigonometric nature of the *mandaphala* is doubtless the reason that Bhāskara chose this point in the *Karaṇakutūhala* to list the successive differences of his *R* sine values and rules for linear interpolation between them. The verses in question, 2.6cd–8 (Mishra, 1991, pp. 21–22), are not quoted here, but the stated *R* sine-differences for each 10° interval in the first quadrant, along with their corresponding cumulative *R* sine values, are reproduced in Schema 18. They enable the user to determine the *R* sine of the *manda*-anomaly as prescribed in *Karaṇakutūhala* 2.9–10 (Mishra, 1991, p. 23):

sūryādikānāṃ mṛdukendradorjyā
digghnī vibhājyātha khapañcabāṇaiḥ ||
nāgāgnidasrair giripūrṇacandrair
vasvaṅkabhūbhir vasunetranetraiḥ || 2.9 || *indravajrā*

yugāṣṭaśailair munipañcacandraiḥ
phalaṃ lavāḥ kendravaśād dhanarṇam ||
āryaṃ grahe sūryavidhū sphuṭau sto *upajāti*
mandasphuṭākhyā itare syur evam || 2.10 || (*premā*)

VERSES 2.9–10

The *R* sines of the *manda*-anomalies of [the planets] beginning with the Sun are multiplied by 10 and [respectively] divided by 550, 238, 107, 198, 228, 784, [and] 157. The result is [in] degrees, positive or negative according to [the value of] the anomaly. [This being] done, [the positions of] the Sun and Moon are corrected; [those of] the other planets are called '*manda*-corrected' in this way.

This algorithm is an approximation to the trigonometrically exact formula for the *manda*-equation discussed above. In it, the amount of the planet's *manda*-equation μ is assumed to vary linearly with the *R* sine of the *manda*-anomaly κ_M (so its values are symmetric about an absolute maximum value that occurs when the anomaly is either 90° or 270°):

$$\mu \approx \frac{10\, R\, \sin \kappa_M}{\text{divisor}}.$$

The table in Schema 19 shows the planets' maximum *manda*-equations reconstructed using the formulae thus specified. For comparison, it also lists the corresponding values of maximum μ derived from the trigonometrically exact relation (using the *Siddhāntaśiromaṇi*'s *manda*-parameters and more accurate Sine-table with $R = 3438$).

Bhāskara continues the *manda* procedures in the next two verses of the *Karaṇakutūhala* (Mishra, 1991, p. 24):

bhānoḥ phalaṃ bhair vihṛtaṃ ca candre
madhye vidheyaṃ ravivad dhanarṇam ||
svabhogyakhaṇḍaṃ navahṛt kharāṃśor
viśvāhataṃ vedahṛtaṃ himāṃśoḥ || 2.11 || *indravajrā*

SCHEMA 19 Parameters and maxima for the *manda*-correction to planetary mean longitude in the *Siddhāntaśiromaṇi* and *Karaṇakutūhala*. Column 2: The planet's *manda*-epicycle circumference, in arclength-units of which there are 360 in a circle of radius R, as stated in *Siddhāntaśiromaṇi* 2.22 (Śāstrī, 1989, p. 45). Column 3: The planet's *manda*-eccentricity ratio implied by the epicycle size in column 2. Column 4: Maximum μ value derived trigonometrically from the *manda*-epicycle parameter. Column 5: Expression for maximum μ value as prescribed in *Karaṇakutūhala* 2.9–10. Column 6: Maximum μ value derived from the expression in column 5

Planet	*manda*-circ.	$\longrightarrow r_M/R$	$\longrightarrow$ max. $\mu^{(\circ)}$	$R \cdot 10/$divisor	$\longrightarrow$ max. $\mu^{(\circ)}$
Sun	13;40	41/1080	2; 10, 31, 0	120 · 10 / 550	2; 10, 54, 32
Moon	31;36	79/900	5; 2, 26, 33	120 · 10 / 238	5; 2, 31, 15
Mars	70	7/36	11; 12, 27, 58	120 · 10 / 107	11; 12, 53, 49
Mercury	38	19/180	6; 3, 30, 56	120 · 10 / 198	6; 3, 38, 10
Jupiter	33	11/120	5; 15, 33, 8	120 · 10 / 228	5; 15, 47, 22
Venus	11	11/360	1; 45, 3, 0	120 · 10 / 784	1; 31, 50, 12
Saturn	50	5/36	7; 58, 53, 6	120 · 10 / 157	7; 38, 35, 55

dvighnaṃ nagāptaṃ kujasaumyayoś ca
khākṣair inaiḥ khārkamitaiś ca bhaktam ||
jīvādikānāṃ ca gateḥ phalaṃ tat-
svarṇaṃ kramāt karkamṛgādikendre || 2.12 || *indravajrā*

VERSES 2.11–2.12

And the equation of the Sun, divided by 27, is to be applied to the mean [longitude of the] Moon, positively or negatively according to [the sign of the equation of] the Sun.

Its own incomplete [R sine-]difference [for the arc of anomaly] is divided by 9 [in the case] of the Sun; multiplied by 13 [and] divided by 4 [in the case] of the Moon. [That quantity in the case] of Mars and Mercury is multiplied by 2 [and] divided by 7, and [in the case] of Jupiter and the rest, divided by 50, 12, and 120 [respectively]. That is the correction of the [daily] velocity, [applied] positively or negatively when the anomaly [is in the half-circle] beginning with Cancer or Capricorn, respectively.

The first half-verse in this excerpt states a rule accounting for the effect upon the Moon's mean longitude of the component of the equation of time caused by the *manda*-equation of the Sun. Since this correction is apparently disregarded in the *Brahmatulyasāraṇī*, we do not analyse it here.

The remaining lines prescribe formulae for a *manda*-correction $\Delta\nu_M$ to planetary velocity in arcminutes per day. These formulae relate each planet's mean velocity $\bar{\nu}$ to its *manda*-corrected version ν_M by means of a planet-specific scale factor, as follows:

$$\nu_M = \bar{\nu} \pm \Delta\nu_M = \bar{\nu} \pm R\,\text{sine-difference}(\varkappa_M) \cdot \text{scale factor}.$$

This correction term $\Delta\nu_M$ is necessary because a planet's *manda*-anomaly affects not only its position but also its speed: its corrected angular velocity is the rate of change of its corrected longitude. The rate of change would be constant were it not for the non-uniformly varying *manda*-equation dependent on the *manda*-anomaly.

SCHEMA 20 Parameters and maxima for the *manda*-correction to planetary mean velocity in the *Siddhāntaśiromaṇi* and *Karaṇakutūhala*. Column 2: The daily rate of change, in degrees, arcminutes and arcseconds, of the planet's *manda*-anomaly (which, for all planets with an effectively stationary *manda*-apogee, is just the planet's own mean daily velocity). Column 3: The velocity-correction scale factor prescribed for the planet in *Karaṇakutūhala* 2.11cd–12. Column 4: The planet's absolute maximum velocity correction term $\Delta\nu_M$, in arcminutes, reconstructed from the formula in *Siddhāntaśiromaṇi* 2.36–38. Column 5: The planet's absolute maximum velocity correction term $\Delta\nu_M$, in arcminutes, reconstructed from the formula in *Karaṇakutūhala* 2.11cd–12

Planet	$(\bar{\nu} - \nu_M)^{(\circ/d)}$	Scale factor	Max. $\Delta\nu_M{}^{(\prime)}$ (*SŚi*)	Max. $\Delta\nu_M{}^{(\prime)}$ (*KKu*)
Sun	0; 59, 8	1/9	2; 14, 41	2; 20, 0
Moon	13; 3, 53	13/4	1, 8; 48, 32	1, 8; 15, 0
Mars	0; 31, 26	2/7	6; 6, 43	6; 0, 0
Mercury	0; 59, 8	2/7	6; 14, 30	6; 0, 0
Jupiter	0; 4, 59	1/50	0; 27, 24	0; 25, 12
Venus	0; 59, 8	1/12	1; 48, 24	1; 45, 0
Saturn	0; 2, 0	1/120	0; 16, 40	0; 10, 30

In *Siddhāntaśiromaṇi* 2.36–38 (Śāstrī, 1989, pp. 52–53), Bhāskara defines a related form of $\Delta\nu_M$ by a rule equivalent to the following equation:

$$\nu_M = \bar{\nu} \pm \Delta\nu_M = \bar{\nu} \pm \frac{(\bar{\nu} - \nu_{A_M}) \cdot R\cos\varkappa_M \cdot (r_M/R)}{R},$$

where ν_{A_M} is the velocity of the planet's *manda*-apogee (which, again, is considered zero except in the case of the Moon). For Bhāskara's own elegant demonstration of this rule in the *Siddhāntaśiromaṇi*, *vide* (Rao, 2017). Here we merely justify it briefly in modern mathematical terms, assuming that the exact trigonometric expression for $\sin\mu$ is more or less equivalent to the following approximation (taking $\mu \approx \sin\mu$ and the *manda*-hypotenuse nearly equal to R):

$$\mu \approx \frac{R\sin\varkappa_M \cdot (r_M/R)}{R}.$$

If we use this approximation in differentiating with respect to time the basic formula for the *manda*-corrected mean longitude λ_M (recalling that the derivative of a longitude is a velocity), we arrive at the following relation:

$$\begin{aligned}
\lambda_M &= \bar{\lambda} \pm \mu, \\
\frac{d}{dt}(\lambda_M) &= \frac{d}{dt}(\bar{\lambda}) \pm \frac{d}{dt}(\mu), \\
\nu_M &= \bar{\nu} \pm \frac{d}{dt}(\mu) \\
&= \bar{\nu} \pm \frac{d}{d\varkappa_M}(\mu) \cdot \frac{d}{dt}(\varkappa_M) \\
&\approx \bar{\nu} \pm \frac{R\cos\varkappa_M \cdot (r_M/R)}{R} \cdot (\bar{\nu} - \nu_{A_M}),
\end{aligned}$$

which is the *Siddhāntaśiromaṇi* formula for the *manda*-corrected mean velocity.

The approximate formula stated in *Karaṇakutūhala* 2.11cd–12 replaces the above correction term $\Delta\nu_M$ for a given planet by a scale factor multiplied by the 'incomplete [R sine-]difference' of the planet's *manda*-anomaly: i.e., whichever one of the nine tabulated R sine-differences (*vide* Schema 18) corresponds to the 10° interval within which the planet's *manda*-anomaly $\varkappa_M$ falls. (Note that this R sine-difference very roughly approximates the R cosine in the *Siddhāntaśiromaṇi* formula. The maximum velocity corrections resulting from the exact and approximate formulae, along with their respective scale factors, are listed in Schema 20.) Since the *Karaṇakutūhala* rule does not call for interpolating within that 10° interval, its version of the velocity correction is a step function rather than a continuous one, as illustrated in figure 8 in section 3.4.

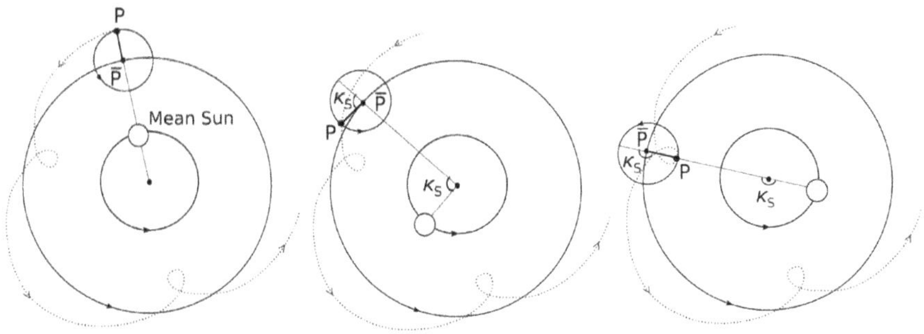

FIGURE 16 The *śīghra*-correction for a superior planet. **Left:** The *śīghra*-anomaly $\varkappa_S$ and the corresponding correction are zero when the planet is in conjunction with the Sun. **Center:** The direction of the planet stays parallel to that of the Sun, which is revolving faster than the mean planet on its orbit, so the planet appears to slow down in its forward motion. **Right:** The continued motion of the mean Sun appears to drag the planet backwards, so that it reaches the center of its retrograde motion in opposition to the Sun, with anomaly 180°.

2.8 *The* śīghra-*Correction to* manda-*Corrected Longitude*

As explained in Appendix, section 2.5, the concept used in Indian astronomy to model a star-planet's synodic inequality is the 'fast', or *śīghra*, anomaly $\varkappa_S$, measured from a notional point called the *śīghra*-apogee. This apogee's position coincides with that of the mean Sun in the case of superior planets. For the inferior planets whose own mean longitude equals that of the mean Sun, the *śīghra*-apogee is tabulated as a separate body, as described in Appendix, section 2.2.

The *śīghra*-anomaly $\varkappa_S$ is determined by subtracting the planet's longitude corrected by the *manda*-equation, or λ_M, from that of its *śīghra*-apogee, λ_{A_S}. Since the *śīghra*-apogee revolves about the earth faster than the mean planet does, the planet periodically appears to go backwards while it is close to its opposition (or in the case of an inferior planet, its inferior conjunction) with respect to the Sun. Figure 16 qualitatively illustrates the *śīghra* for a superior planet, neglecting the effect of the *manda*: as the *śīghra*-anomaly angle goes from 0° to 180°, the planet's apparent motion follows the dotted path, whose loops represent the apparent retrogradations.

This cyclic 'looping' means that the values of the *śīghra*-equation σ are symmetric about the end of the second quadrant of anomaly. For instance, σ is zero when the anomaly is zero (at conjunction or superior conjunction for a superior or inferior planet respectively) or 180° (opposition/inferior conjunction); it is added to the *manda*-corrected longitude λ_M in the first two quadrants of anomaly, and subtracted from λ_M in the second two. More precisely, its value at an arbitrary $\varkappa_S$ is determined by the formula stated in *Karaṇakutūhala* 2.13 (Mishra, 1991, p. 25):

koṭijyā calakendrajā paraguṇā dvighnī tayonānvitā
kendre karkamṛgādike parakṛtiḥ svābhrābdhiśakrair yutā ||
tanmūlaṃ śravaṇaḥ pareṇa guṇitā dorjyātha karṇoddhṛtāḥ
taccāpaṃ capalaṃ phalaṃ dhanam ṛṇaṃ mandasphuṭe
syāt sphuṭaḥ || 2.13 || *śārdūlavikrīḍita*

VERSE 2.13

The R cosine produced from the *śīghra*-anomaly is multiplied by the *śīghra*-radius [and] multiplied by 2. The square of the *śīghra*-radius is diminished or increased by that, when the anomaly is [in the half-circle] beginning with Cancer or Capricorn [respectively], and added to 14400 [i.e., $120^2 = R^2$]. The square-root of that is the hypotenuse. Now the R sine [of the *śīghra*-anomaly] is multiplied by the *śīghra*-radius [and] divided by the hypotenuse. The arc from that is the equation [of the] *śīghra*[-anomaly]. [When it is applied] positively or negatively to the *manda*-corrected [longitude of the planet, the longitude] should be correct.

In modern notation, this algorithm is equivalent to the equation

$$R\sin\sigma = \frac{R\sin\varkappa_S \cdot r_S}{\sqrt{{r_S}^2 \pm 2\,r_S\,R\cos\varkappa_S + R^2}} = \frac{R \cdot r_S\sin\varkappa_S}{H_S},$$

where r_S is the *para* or radius of the planet's *śīghra*-epicycle (see Schema 17) and the so-called *śīghra*-hypotenuse H_S extends from the earth to the planet's true position. Note that this equation can trivially be rewritten in a form analogous to that for the trigonometrically exact *manda*-correction stated in Appendix, section 2.7:

$$\sin\sigma = \frac{R\sin\varkappa_S \cdot \frac{r_S}{R}}{\sqrt{\left(R\sin\varkappa_S \cdot \frac{r_S}{R}\right)^2 + \left(R \pm R\cos\varkappa_S \cdot \frac{r_S}{R}\right)^2}}.$$

Both are equivalent to Bhāskara's prescription for σ in *Siddhāntaśiromaṇi* 2.32–33 (Śāstrī, 1989, pp. 50–51). The large size of the *śīghra*-radius here, and the correspondingly large variation in σ, preclude the use of a simpler approximate formula as in the *manda*-equation rule of *Karaṇakutūhala* 2.9–10.

To determine analytically where the maximum σ-values will occur, we note that qualitatively, their derivation exactly parallels the geometric rationale for the location of the theoretical maximum *manda*-equation that we gave in Appendix, section 2.7: namely,

$$\varkappa_S = 90^\circ + \arcsin\left(\frac{r_S}{R}\right) \quad \text{or} \quad 270^\circ - \arcsin\left(\frac{r_S}{R}\right).$$

Alternatively, we can set the derivative of $\sigma = \arcsin(\sin\sigma)$ to zero and solve for the σ-maximizing value(s) of $\varkappa_S$ (using modern sine and cosine functions generalized to any quadrant):

$$\frac{d}{d\varkappa_S}(\arcsin(\sin\sigma)) = \frac{d}{d\varkappa_S}\left(\arcsin\left(\frac{r_S \sin\varkappa_S}{\sqrt{{r_S}^2 + 2Rr_S\cos\varkappa_S + R^2}}\right)\right)$$

$$= \frac{d}{d\varkappa_S}\left(\arcsin\left(\frac{r_S\sin\varkappa_S}{H_S}\right)\right)$$

$$= \frac{1}{\sqrt{1 - \frac{{r_S}^2\sin^2\varkappa_S}{{H_S}^2}}} \cdot \frac{(r_S\cos\varkappa_S)(H_S) - \frac{r_S\sin\varkappa_S(-2Rr_S\sin\varkappa_S)}{2H_S}}{{H_S}^2}$$

$$= \left(\frac{1}{\sqrt{1 - \frac{{r_S}^2\sin^2\varkappa_S}{{H_S}^2}}}\right) \cdot \left(\frac{(r_S\cos\varkappa_S)\sqrt{{r_S}^2 + 2Rr_S\cos\varkappa_S + R^2} - \frac{r_S\sin\varkappa_S(-2Rr_S\sin\varkappa_S)}{2\sqrt{{r_S}^2 + 2Rr_S\cos\varkappa_S + R^2}}}{{H_S}^2}\right)$$

$$= \frac{1}{\sqrt{\frac{{H_S}^2 - {r_S}^2\sin^2\varkappa_S}{{H_S}^2}}} \cdot \frac{r_S\cos\varkappa_S({H_S}^2) + R{r_S}^2\sin^2\varkappa_S}{({H_S}^2)H_S}$$

$$= \left(\frac{H_S}{\sqrt{{r_S}^2\cos^2\varkappa_S + 2r_S\cos\varkappa_S + R^2}}\right) \cdot \left(\frac{{r_S}^3\cos\varkappa_S + 2R{r_S}^2\cos^2\varkappa_S + R^2 r_S\cos\varkappa_S + R{r_S}^2\sin^2\varkappa_S}{({H_S}^2)H_S}\right)$$

$$= \frac{{r_S}^3 \cos \varkappa_S + R {r_S}^2 \cos^2 \varkappa_S + R^2 r_S \cos \varkappa_S + R {r_S}^2}{\sqrt{(r_S \cos \varkappa_S + R)^2 \cdot ({H_S}^2)}}$$

$$= \frac{(r_S \cos \varkappa_S + R)({r_S}^2 + R r_S \cos \varkappa_S)}{\sqrt{(r_S \cos \varkappa_S + R)^2 \cdot ({r_S}^2 + 2 R r_S \cos \varkappa_S + R^2)}} = \frac{{r_S}^2 + R r_S \cos \varkappa_S}{{r_S}^2 + 2 R r_S \cos \varkappa_S + R^2}.$$

When σ is at its maximum, this reduces to

$$\begin{aligned} 0 &= \frac{{r_S}^2 + R r_S \cos \varkappa_S}{{r_S}^2 + 2 R r_S \cos \varkappa_S + R^2} \\ &= {r_S}^2 + R r_S \cos \varkappa_S \\ &= r_S + R \cos \varkappa_S, \end{aligned}$$

$$\cos \varkappa_S = -\frac{r_S}{R},$$

or equivalently,

$$\begin{aligned} \varkappa_S &= \arccos\left(-\frac{r_S}{R}\right) \\ &= 90^\circ + \arcsin\left(\frac{r_S}{R}\right) \quad \text{or} \quad 270^\circ - \arcsin\left(\frac{r_S}{R}\right). \end{aligned}$$

2.9 *Iteration of the Longitude Corrections*

When a planet's *śīghra*-equation σ is applied to its *manda*-corrected mean longitude λ_M, the result is the so-called 'true longitude' λ:

$$\lambda = \lambda_M \pm \sigma.$$

However, the sequential application of the *manda* and *śīghra* corrections separately somewhat misrepresents their simultaneous influence. Presumably in consequence of this, Sanskrit treatises on spherical astronomy typically direct the user to iterate the sequence of corrections, either for a specified number of steps or until the resulting true longitude becomes fixed; *vide* (Duke, 2005) for a detailed analysis of such procedures. *Karaṇakutūhala* 2.14 (Mishra, 1991, p. 26) prescribes initially halving both equations in the case of Mars:

taduttham adyena calena madhyaś
cet saṃskṛtaṃ spaṣṭataras tadā syāt ||
dalīkṛtābhyāṃ prathamaṃ phalābhyāṃ *upajāti*
tato 'khilābhyām asakṛt kujas tu || 2.14 || (*ṛddhi*)

VERSE 2.14

When the mean [longitude] is corrected [again] by the *manda*[-equation] arising from that [previous correction, and then by the subsequent] *śīghra*[-equation], then it should be more accurate. But [the longitude of] Mars [should be corrected] first by both equations halved and then by both entire [equations] again.

2.10 *The* śīghra-*Correction to* manda-*Corrected Velocity*

As in the case of the *manda*-correction, the *śīghra*-equation affects not only the planet's longitude but also its velocity. *Karaṇakutūhala* 2.16 (Mishra, 1991, p. 27)[14] states the formula for this velocity correction:

drākkendrabhuktir guṇitāśu cāpa-
bhogajyayā khābdhi guṇā ca bhaktā ‖
saptaghnakarṇena caloccabhukteḥ
śodhyā viśeṣā sphuṭakheṭabhuktiḥ ‖ 2.16 ‖ *indravajrā*

VERSE 2.16

The velocity of the *śīghra*-anomaly is multiplied by the incomplete [*R* sine]-difference of the arc of the *śīghra*-[equation], and multiplied by 40, and divided by the hypotenuse multiplied by 7. [That] is to be subtracted from the velocity of the *śīghra*-apogee. The remainder is the corrected velocity of the planet.

We can understand the 'velocity of the *śīghra*-anomaly' or v_{κ_S} as the difference between the velocity v_{A_S} of the *śīghra*-apogee (which, unlike that of the

14 The text of the *Karaṇakutūhala* here diverges somewhat between the edition in (Mishra, 1991), (Rao and Uma, 2008), and the manuscripts edited in (Plofker, n.d.). The former omits the half-verse that we denote 2.17ab, and includes an extra verse (quoted below) that appears to explain the meaning of '*manda*-corrected velocity' and '*śīghra*-anomaly velocity':
gateḥ phalenaiva tu saṃskṛtā yā
mudhyā gatir mandagatir bhavet sā ‖
grahasya mandasphuṭabhuktivarjitā
svā śīghrakendrasya gatir bhavet sā ‖ 2.15 ‖
'Whatever is the mean velocity corrected by the [*manda*-]correction of the velocity, that is the *manda*-corrected velocity. Its own [i.e., the *śīghra*-apogee velocity?] decreased by the *manda*-corrected velocity of the planet is the velocity of the *śīghra*-anomaly.'

much slower *manda*-apogees, cannot be assumed to be zero) and the planet's own *manda*-corrected velocity ν_M. The above rule for 'corrected velocity' ν in arcminutes per day therefore becomes equivalent to the following expression:

$$\nu = \nu_{A_S} - \frac{\nu_{\kappa_S} \cdot R\,\text{sine-difference}(\sigma)}{H_S} \cdot \frac{40}{7},$$

where H_S as before denotes the *śīghra*-hypotenuse. Compare Bhāskara's corresponding rule for true velocity ν in *Siddhāntaśiromaṇi* 2.39 (Śāstrī, 1989, p. 54), which prescribes:

$$\nu = \nu_{A_S} - \frac{\nu_{\kappa_S} \cdot R\cos\sigma}{H_S}.$$

The *Karaṇakutūhala* algorithm, like its counterpart for the *manda*-derived velocity correction discussed in Appendix, section 2.7, merely replaces the R cosine in its *Siddhāntaśiromaṇi* counterpart by the appropriately scaled R sine-difference for the 10° interval in which σ falls, thus:

$$R\cos\sigma \approx \frac{R\,\text{sine-diff.}(\sigma)\cdot R}{\text{max.}\ R\,\text{sine-diff.}} = \frac{R\,\text{sine-diff.}(\sigma)\cdot 120}{21} = \frac{R\,\text{sine-diff.}(\sigma)\cdot 40}{7}.$$

The results of the two formulae in the case of Jupiter are illustrated by the graph in figure 17.[15]

A modern mathematical rationale for Bhāskara's exact true-velocity formula can be derived by considering each of the velocities ν_{A_S} and ν_M as simply a change between two successive positions in longitude with superscripts *i* and *ii*:

$$\begin{aligned}\left(\lambda_{A_S}{}^{ii} - \lambda_M{}^{ii}\right) - \left(\lambda_{A_S}{}^{i} - \lambda_M{}^{i}\right) &= \left(\lambda_{A_S}{}^{ii} - \lambda_{A_S}{}^{i}\right) - \left(\lambda_M{}^{ii} - \lambda_M{}^{i}\right)\\ &= \nu_{A_S} - \nu_M\\ &= \nu_{\kappa_S}.\end{aligned}$$

15 Note that a slightly modified version of this figure in (Montelle and Plofker, 2015, p. 25) misleadingly describes the pictured functions as '*śīghra* velocity correction' instead of '*śīghra*-corrected velocity'.

And since the true planetary longitude λ is given by

$$\lambda = \lambda_M + \sigma$$

(where σ is considered a signed quantity), and we can regard velocity as just the derivative of longitude with respect to time, the true velocity ν can be defined using the rule for σ from Appendix, section 2.8:

$$\begin{aligned} \nu &= \frac{d}{dt}(\lambda) = \frac{d}{dt}(\lambda_M + \sigma) = \frac{d}{dt}(\lambda_M) + \frac{d}{dt}(\sigma) \\ &= \frac{d}{dt}(\lambda_M) + \frac{d}{dt}\left(\arcsin\left(\frac{r_S \sin \varkappa_S}{H_S}\right)\right) \\ &= \frac{d}{dt}(\lambda_M) + \frac{d}{d\varkappa_S}\left(\arcsin\left(\frac{r_S \sin \varkappa_S}{H_S}\right)\right) \cdot \frac{d}{dt}(\varkappa_S) \\ &= \nu_M + \frac{d}{d\varkappa_S}(\sigma) \cdot \nu_{\varkappa_S}. \end{aligned}$$

Substituting $\nu_{A_S} - \nu_{\varkappa_S}$ for ν_M and recalling our result for the derivative of σ from Appendix, section 2.8, we arrive at

$$\begin{aligned} \nu &= \nu_{A_S} - \nu_{\varkappa_S} \cdot \left(1 - \frac{d}{d\varkappa_S}(\sigma)\right) \\ &= \nu_{A_S} - \nu_{\varkappa_S} \cdot \left(1 - \frac{{r_S}^2 + R\, r_S \cos \varkappa_S}{{r_S}^2 + 2\, R\, r_S \cos \varkappa_S + R^2}\right) \\ &= \nu_{A_S} - \nu_{\varkappa_S} \cdot \left(1 - \frac{{r_S}^2 + 2\, R\, r_S \cos \varkappa_S + R^2 - (R\, r_S \cos \varkappa_S + R^2)}{{r_S}^2 + 2\, R\, r_S \cos \varkappa_S + R^2}\right) \\ &= \nu_{A_S} - \nu_{\varkappa_S} \cdot \left(1 - \frac{{H_S}^2 - (R\, r_S \cos \varkappa_S + R^2)}{{H_S}^2}\right) \\ &= \nu_{A_S} - \nu_{\varkappa_S} \cdot \left(1 - \left(1 - \frac{R\,(R + r_S \cos \varkappa_S)}{{H_S}^2}\right)\right) \\ &= \nu_{A_S} - \nu_{\varkappa_S} \cdot \left(\frac{R}{H_S} \cdot \frac{R + r_S \cos \varkappa_S}{H_S}\right) \\ &= \nu_{A_S} - \nu_{\varkappa_S} \cdot \left(\frac{R}{H_S} \cdot \cos \sigma\right), \end{aligned}$$

which is Bhāskara's rule.

Appendix, section 2.8 briefly explained why the synodic or *śīghra* anomaly causes a star-planet in certain circumstances to appear to be moving backwards in longitude or retrograding. *Karaṇakutūhala* 2.17ab (Plofker, n.d.) states the condition and the required velocity correction for this period of retrogradation:

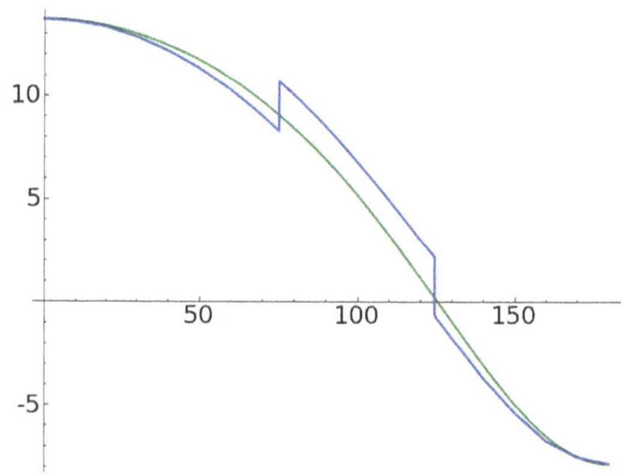

FIGURE 17 The *śīghra*-corrected true velocity in arcminutes per day, assuming $v_M = \bar{v}$, for 0–180° of *śīghra*-anomaly for Jupiter: The *Siddhāntaśiromaṇi* formula (smooth; using modern trig functions) and the *Karaṇakutūhala* approximation to it (piecewise smooth; using trig-function results from interpolating in the *Karaṇakutūhala* *R* sine table)

yadā na śuddhā viparītaśodhyā *upajāti*
śeṣaṃ bhaved vakragatis tadānīm ‖ 2.17ab ‖ (*haṃsī*)

VERSE 2.17ab

When it [i.e., the *śīghra*-anomaly velocity v_{κ_S} modified as directed in verse 2.16] is not [capable of being] subtracted [from the *śīghra*-apogee velocity v_{A_S}], it should be subtracted in reverse; the remainder is the retrograde velocity at that time.

That is, if the modified *śīghra*-anomaly velocity is greater than the velocity of the *śīghra*-apogee, subtracting the former from the latter will produce a negative (i.e., retrograde) true velocity v, whose absolute value is given by subtracting 'in reverse' the latter from the former. The so-called 'stationary points' between direct and retrograde motion are the zeroes of the function(s) for v illustrated in figure 17; they occur symmetrically about 180° of *śīghra*-anomaly.

2.11 *Rising-Difference or* udayāntara *Corrections for the Sun and Moon*

The 'rising-difference' (*udayāntara*) is a time correction equivalent to what modern astronomy considers one of the components of the 'equation of time'. The problem addressed by the equation of time arises from the fact that standard astronomical time units are constant in length, as though measuring the mean Sun rolling uniformly around the celestial equator. But the observed events that regulate timekeeping practice, such as sunrise, sunset and the Sun's culmination at true local noon, all depend instead upon the Sun's true motion, which is irregular due to both its own orbital anomaly and the tilt of the ecliptic with respect to the celestial equator. The effect of this discrepancy upon computed time intervals is broken down into two components:

- **Solar eccentricity**
  Due to the *manda*-correction described in Appendix, section 2.7 as more or less accounting for the eccentricity of the Sun's orbit, the Sun's true ecliptic longitude generally differs by a small amount from its mean longitude. This small longitudinal arc corresponds to some amount of elapsed time.
- **Ecliptic obliquity**
  The tilt angle of the ecliptic to the celestial equator is called the ecliptic obliquity, denoted ϵ, and in Sanskrit astronomy typically assigned the value 24°. Due to this inclination, arcs of equal length on different parts of the ecliptic correspond to different-sized equatorial arcs or 'right ascensions' measured in time-degrees, depending on their positions with respect to the vernal point or intersection of the ecliptic with the equator. *Vide* figure 19, where the vernal point at the beginning of Aries in the tropical zodiac is denoted ♈.

The so called *udayāntara* accounts for the latter discrepancy using the Sun's tropical mean longitude $\bar{\lambda}_\odot$, i.e., its mean ecliptic longitude measured from the vernal point. When $\bar{\lambda}_\odot$ falls in an odd quadrant, it will be greater than its accompanying right ascension arc α by $\bar{\lambda}_\odot - \alpha$ degrees (in an even quadrant, on the other hand, α is greater than $\bar{\lambda}_\odot$). Since one day or 60 *ghaṭikās* corresponds to 360 time-degrees on the equator, the discrepancy in time produced by this rising-difference will be $\frac{\bar{\lambda}_\odot - \alpha}{6}$ *ghaṭikās*. The size of α itself is determined by the spherical-trigonometry relation

$$\sin\alpha = \sin\bar{\lambda}_\odot \cdot \frac{\cos\epsilon}{\cos\delta(\bar{\lambda}_\odot)}$$

where $\delta(\bar{\lambda}_\odot)$ is the ecliptic declination of the point at longitude $\bar{\lambda}_\odot$ (*vide* Appendix, section 2.12). So for a planet moving with mean velocity $\bar{v}$ arcminutes per day or $\bar{v}/60$ arcminutes per *ghaṭikā*, the adjustment to its longitude required by this time discrepancy will be

$$\begin{aligned} udayāntara &= \frac{\bar{\lambda}_{\odot} - \alpha}{6} \cdot \frac{\bar{v}}{60} \text{ arcminutes} \\ &= \frac{\bar{\lambda}_{\odot} - \alpha}{6} \cdot \bar{v} \text{ arcseconds} \\ &= \frac{1}{6} \cdot \left(\bar{\lambda}_{\odot} - \arcsin \left(\sin \bar{\lambda}_{\odot} \cdot \frac{\cos \epsilon}{\cos \delta(\bar{\lambda}_{\odot})} \right) \right) \cdot \bar{v} \text{ arcseconds.} \end{aligned}$$

In *Siddhāntaśiromaṇi* 2.65 (Śāstrī, 1989, p. 65), Bhāskara simplifies the *udayāntara* calculation with an approximation using $R = 120$, as follows:

$$udayāntara \approx \frac{R \sin \left(2 \bar{\lambda}_{\odot} \right)}{270} \cdot \bar{v} \text{ arcseconds.}$$

In his autocommentary (Śāstrī, 1989, pp. 65–66), he points out that the difference between $\bar{\lambda}_{\odot}$ and α will be maximum when $\bar{\lambda}_{\odot}$ is in the middle of a quadrant, and zero at its ends, and thus can be computed by a 'sine technique' (*jyāprakāra*) or linear proportion using $R \sin \left(2 \bar{\lambda}_{\odot} \right)$. This maximum difference is around $2^{\circ}\ 30'$ when $\bar{\lambda}_{\odot}$ equals 45°, at which point $R \sin \left(2 \bar{\lambda}_{\odot} \right) = R = 120$. So it seems plausible that Bhāskara took the factor $120/270$ as sufficiently close to $2.5/6$ and allowed the numerator to vary as the R sine of $2 \bar{\lambda}_{\odot}$.[16] The trigonometrically exact formula and the *Siddhāntaśiromaṇi* formula for the *udayāntara* of the Sun are graphed along with the *Karaṇakutūhala*'s cruder approximation in figure 18.

To find the *udayāntara* by whatever method, it is necessary to know the location of the vernal point, which is determined by the current amount of precession of the equinoxes. The standard value in Indian astronomy for the amount of this precession is one arcminute per year. Bhāskara's rule in *Karaṇakutūhala* 2.17cd (Mishra, 1991, p. 28) takes the accumulated amount of precession at epoch to be 11°, which implies that it would have been zero 660 years before

16 However, it is never safe to assume that Bhāskara's rationales for approximation formulae are more simple than subtle. On closer examination, we find that the trigonometric function expression for $\bar{\lambda}_{\odot} - \alpha$ has a maximum value of about $2^{\circ}\ 35'$ when its derivative with respect to $\bar{\lambda}_{\odot}$ is zero, at approximately $\bar{\lambda}_{\odot} = 46^{\circ}\ 18'$. This maximum difference produces a maximum solar *udayāntara* of about 25;31 arcseconds. Yet Bhāskara's commentary states that the Sun's maximum *udayāntara* is somewhat higher than 26, and that the divisor 270 has been chosen to produce that result when $\bar{\lambda}_{\odot} = 45^{\circ}$. It may be that Bhāskara approximated some quantity or calculation in a way that has not occurred to us.

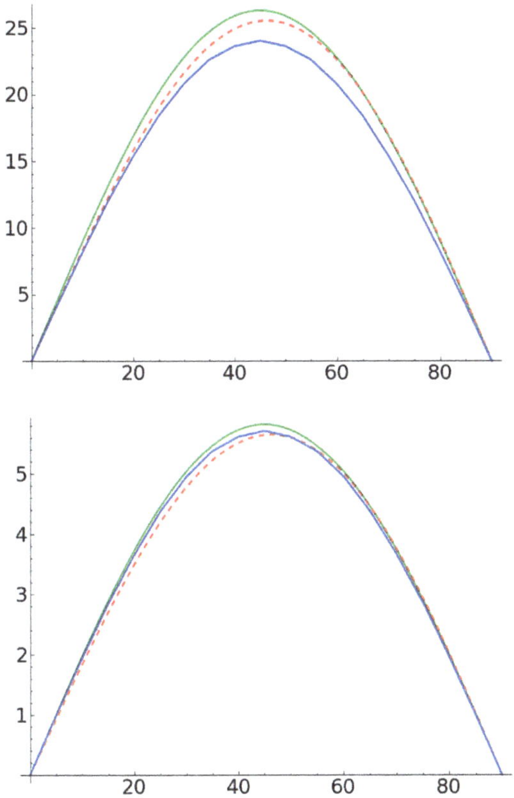

FIGURE 18 Comparison of the rules for the Sun's (above; in arcseconds) and Moon's (below; in arcminutes) *udayāntara* for 0–90° of tropical mean solar longitude $\bar{\lambda}_{\odot}$. **Dashed curve:** Trigonometrically exact formula. **Smooth curve:** Approximation from *Siddhāntaśiromaṇi* 2.65. **Piecewise linear curve:** Simpler approximation from *Karaṇakutūhala* 2.18. The reconstructions used 59′ 8″/day for the mean solar and 790′ 35″/day for the mean lunar velocity.

the epoch of the text, in 523 CE.[17] He then continues in verse 2.18 (Mishra, 1991, p. 29) with the specification of solar and lunar *udayāntara* rules:

athāyanāṃśāḥ karaṇābdaliptā *upajāti*
yuktā bhavās tadyutamadhyabhānoḥ ‖ 2.17cd ‖ (*haṃsī*)

17 See (Pingree, 1972, pp. 28–29), noting the canonical practice of setting the zero of precession at Śaka 444 (522/523 CE).

dighnasya dorjyā śarahṛd viliptā
bhānor vidhoḥ kvakṣihṛtāḥ kalās tāḥ ||
svarṇaṃ ca yugmaujapade sthite 'rke *upajāti*
krameṇa karmety udyāntarākhyam || 2.18 || *(bālā)*

VERSES 2.17cd–18

Now the degrees of precession are 11 plus the minutes [equal to] the years [elapsed since the date] of the handbook [i.e., the *Karaṇakutūhala*]. The *R* sine of the mean [longitude of the] Sun increased by that [and] multiplied by 2, divided by 5, is the [correction in] arcseconds for the Sun. [The same quantity in the case] of the Moon is divided by 21; [the result is in] arcminutes. And these are [applied] positively or negatively when the Sun is standing in an even or odd quadrant respectively. That is the operation called the *udayāntara*.

These rules are evidently simplified forms of the *Siddhāntaśiromaṇi* 2.65 formula discussed above, specialized for the Sun (denoted ☉) and the Moon (denoted ☾):

$$\begin{aligned} \mathit{udayāntara}_{\odot} &\approx \frac{R\sin(2\bar{\lambda}_{\odot})}{5} \qquad \left(\approx \frac{R\sin(2\bar{\lambda}_{\odot})}{270}\cdot 59\right) \text{ arcseconds,} \\ \mathit{udayāntara}_{☾} &\approx \frac{R\sin(2\bar{\lambda}_{\odot})}{21} \qquad \left(\approx \frac{R\sin(2\bar{\lambda}_{\odot})}{270}\cdot \frac{791}{60}\right) \text{ arcminutes.} \end{aligned}$$

The instruction about determining the sign of the *udayāntara* when $\bar{\lambda}_{\odot}$ is in odd or even quadrants conforms to the above remark about the relative sizes of $\bar{\lambda}_{\odot}$ and α.

2.12 *Solar Declination*

The solar declination δ, the ordinate of the right ascension α discussed above, is the angular distance between the equator and the point of the Sun's tropical longitude $\lambda_{\odot}$ on the ecliptic, measured on the great circle passing through this point and the poles of the equator. Its value is determined from $\lambda_{\odot}$ and the ecliptic obliquity ϵ by the trigonometric formula

$$R\sin\delta = \frac{R\sin\lambda_{\odot}\cdot R\sin\epsilon}{R}$$

based on the similar right triangles shown in figure 19.

SCHEMA 21 The δ-difference values in arcminutes stated in *Karaṇakutūhala* 3.13–14 for each 15° of arc in the first quadrant, and their corresponding cumulative δ values

$\lambda_{\odot}$ (°)	δ-difference (′)	δ (′)
15	362	362
30	341	703
45	299	1002
60	236	1238
75	150	1388
90	52	1440

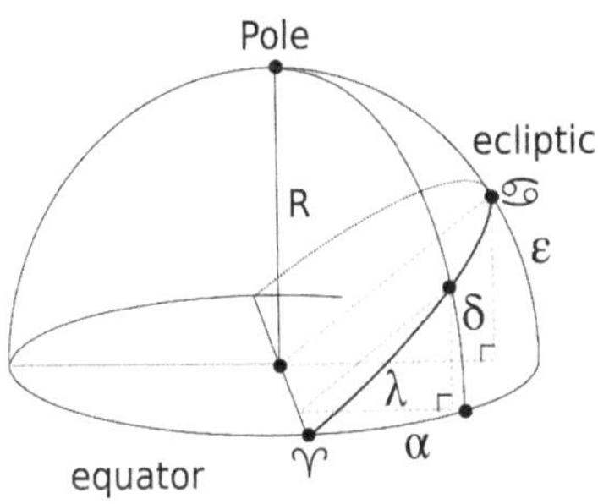

FIGURE 19
The northern celestial hemisphere from the equator to the north celestial pole, showing the northern semicircle of the ecliptic tilted to the equator by the obliquity ϵ, passing through the vernal point ♈ and the summer solstice point ♋. The arc of longitude λ measured from the start of the tropical zodiac at ♈ has corresponding right ascension α and ecliptic declination δ. $R \sin \lambda$ and R are the hypotenuses of the similar right triangles in the sphere with vertical legs $R \sin \delta$ and $R \sin \epsilon$, respectively.

The *Karaṇakutūhala*'s treatment of declination is presented near the end of its third chapter on determining direction, place and time for astronomical observations at a given terrestrial location. Verses 3.13–14 (Mishra, 1991, p. 43) list pre-computed declination values at intervals of 15° in the first quadrant (*vide* Schema 21), while verse 3.15 (Mishra, 1991, p. 44) states an algebraic approximation formula to find δ for an arbitrary $\lambda_{\odot}$ without interpolation. Both methods use the standard Indian value for maximum δ (i.e., ϵ), namely $1440' = 24°$; but since the *Brahmatulyasāraṇī* apparently does not make use of them (*vide* section 3.6.2), we do not discuss them in detail here.

References

Apte, Vaman Shivram (1890). *The Practical Sanskrit-English Dictionary*. Poona: Shiralkar & Co.

Arkasomayaji, D., ed. (1980). *Siddhāntaśiromaṇi of Bhāskarācārya*. Kendriya Sanskrit Vidyapeetha, Tirupati Series no. 29. New Delhi: Rashtriya Sanskrit Sansthan.

Bardi, Alberto (2018). "The Paradosis of the Persian Tables, a Source on Astronomy between the Ilkhanate and the Eastern Roman Empire". In: *Journal for the History of Astronomy* 49.2, pp. 239–260.

Bhandarkar Oriental Research Institute (1990–1991). *Descriptive Catalogue of Manuscripts in the Government Manuscripts Library deposited at the Bhandarkar Oriental Research Institute*. Vol. 30–32. "Vedāṅgas". Poona: Bhandarkar Oriental Research Institute.

Bjørnbo, Axel Anthon, Rasmus Olsen Besthorn, and Heinrich Suter, eds. (1914). *Die astronomischen Tafeln des Muḥammed ibn Mūsā al-Khwārizmī in der Bearbeitung des Maslama ibn Aḥmed al-Madjrīṭī und der latein. Übersetzung des Athelhard von Bath auf Grund der Vorarbeiten von A. Bjørnbo und R. Besthorn in Kopenhagen*. XXV, 255 S.; 4°: graph. Darst. København: Høst.

Burnard, Lou, Katherine O'Brien O'Keeffe, and John Unsworth, eds. (2006). *Electronic Textual Editing*. New York: Modern Language Association of America.

Carter, M.G. (1995). "Arabic Literature". In: *Scholarly Editing: A Guide to Research*. Ed. by D.C. Greetham. New York: Modern Language Association of America, pp. 546–574.

Chabás, José and Bernard R. Goldstein (2012). *A Survey of European Astronomical Tables in the Late Middle Ages*. Leiden: Brill.

Chemla, Karine (2016a). "Numerical Tables in Chinese Writings Devoted to Mathematics: From Early Imperial Manuscripts to Printed Song-Yuan Books". In: *East Asian Science, Technology, and Medicine* 44, pp. 69–121.

Chemla, Karine (2016b). "Special Issue on Numerical Tables and Tabular Layouts in Chinese Scholarly Documents—Part I: On the Work to Produce Tables and the Meaning of their Format". In: *East Asian Science, Technology, and Medicine* 43, pp. 9–15.

Chemla, Karine (2016c). "Special Issue on Numerical Tables and Tabular Layouts in Chinese Scholarly Documents—Part II: Synchronic and Diachronic Approaches to the Texts of Tables". In: *East Asian Science, Technology, and Medicine* 44, pp. 11–20.

Cullen, Christopher (2011). "Understanding the Planets in Ancient China: Prediction and Divination in the 'Wu xing zhan'". In: *Early Science and Medicine* 16.3, pp. 218–251.

Cullen, Christopher (2016). *The Foundations of Celestial Reckoning: Three Ancient Chinese Astronomical Systems*. Scientific Writings from the Ancient and Medieval World. Abingdon: Routledge.

Driscoll, Matthew James and Elena Pierazzo, eds. (2016). *Digital Scholarly Editing: Theories and Practices*. Digital Humanities Series 4. Cambridge: Open Book Publishers.

Duke, Dennis (2005). "The Equant in India: the Mathematical Basis of Indian Planetary Models". In: *Archive for History of Exact Sciences* 59, pp. 563–576.

Dvivedī, Sudhākara, ed. (1901–1902). *Brāhmasphuṭasiddhānta*. The Pandit NS 23–24. Benares: Government Sanskrit College.

Gokhale, Shobhana (1967). *Indian Numerals*. Poona: Deccan College Postgraduate and Research Institute.

González-Reimann, Luis (2009). "Cosmic Cycles, Cosmology, and Cosmography". In: *Brill's Encyclopedia of Hinduism*. Ed. by Knut A. Jacobsen. Vol. 1. Leiden: Brill, pp. 411–428.

Greetham, D.C. (1993). "Editorial and Critical Theory from Modernism to Postmodernism". In: *Palimpsest: Editorial Theory in the Humanities*. Ed. by George Bornstein and Ralph G. Williams. Ann Arbor: University of Michigan Press, pp. 9–28.

HAMSI (2017). *History of Astronomical and Mathematical Sciences in India*. http://www.hamsi.org.nz/.

Hasse, Dag Nikolaus, David Juste, and Benno van Dalen (2013). *Ptolemaeus Arabus et Latinus*. https://ptolemaeus.badw.de/start.

Hayashi, Takao, Clemency Montelle, and K. Ramasubramanian, eds. (2019). *Bhāskaraprabhā*. Culture and History of Mathematics 11. New Delhi: Hindustan Book Agency.

Husson, Matthieu (2015). *TAMAS in brief: how and why care about astronomical tables?* https://f-origin.hypotheses.org/wp-content/blogs.dir/2974/files/2015/12/TAMAS.pdf.

Husson, Matthieu (2018). *Digital Information Systems in the History of Astral Sciences*. https://dishas.obspm.fr.

Husson, Matthieu (2017). *Alfonsine Astronomy*. https://alfa.hypotheses.org/about.

Jones, Alexander (1999). *Astronomical Papyri from Oxyrhynchus*. Philadelphia: American Philosophical Society.

Katre, S.M. (1954). *Introduction to Indian Textual Criticism*. Pune: Deccan College Postgraduate and Research Institute.

Kennedy, E.S. (1956). "A Survey of Islamic Astronomical Tables". In: *Transactions of the American Philosophical Society New Series* 46.2, pp. 123–177.

King, David A. (2004). *In Synchrony with the Heavens: Studies in Astronomical Timekeeping and Instrumentation in Medieval Islamic Civilization*. Leiden and Boston: Brill.

King, David A. and Julio Samsó (2001). "Astronomical Handbooks and Tables from the Islamic World (750–1900): An Interim Report". In: *Suhayl* 2, pp. 9–105.

Kolachana, Aditya et al. (2018). "The *Candrārkī* of Dinakara: A Text Related to Solar and Lunar Tables". In: *Journal for the History of Astronomy* 49.3, pp. 306–344.

Lancaster, Lewis (n.d.). *Critical Editing of Buddhist Texts*. http://www.academia.edu/7284930/Critical_Editing_of_Buddhist_Texts.

Li, Liang (2014). "Tables with 'European' Layout in China: A Case Study in Tabular Layout Transmission". In: *Suhayl* 13, pp. 83–101.

Li, Liang (2016). "Arabic Astronomical Tables in China: Tabular Layout and its Implications for the Transmission and Use of the Huihui lifa". In: *East Asian Science, Technology, and Medicine* 44, pp. 21–68.

Maas, Paul (1956). *Textual Criticism*. Oxford: Oxford University Press.

Makaranda (1923). *Makarandasāraṇī*. Bombay: Sree Venkatesware Press. https://archive.org/details/in.ernet.dli.2015.405753.

Mercier, Raymond (1994). *An Almanac for Trebizond for the Year 1336*. Corpus des astronomes byzantins 7. Louvain-la-Neuve: Institut Orientaliste de Louvain.

Mercier, Raymond (2011). *Les Tables Faciles de Ptolémée: 1b Tables A1–A2. Transcription and Commentary*. Louvain-La-Neuve: Université Catholique de Louvain/Peeters.

Mishra, Satyendra, ed. (1991). *Karaṇakutūhala of Bhāskarācārya*. Krishnadas Sanskrit Series 129. Varanasi: Krishnadas Academy.

Misra, Anuj (2016). "The *Golādhyāya* of Nityānanda's *Sarvasiddhāntarāja*: An Examination of 'The Chapter on Spheres' in a Seventeenth Century Text on Mathematical Astronomy". PhD thesis. University of Canterbury.

Misra, Anuj, Clemency Montelle, and Kim Plofker (2016). "Eclipse Computation Tables in Sanskrit Astronomy: A Critical Edition of the Tables of the *Karaṇakesarī* of Bhāskara (fl. c. 1681)". In: *History of Science in South Asia* 4, pp. 1–79.

Montelle, Clemency (2014). "The Emergence of 'Cyclic' Tables in Indian Astronomy in the Seventeenth Century: Haridatta's *Jagadbhūṣaṇa* and its Islamic Inspiration". In: *Suhayl* 13, pp. 63–81.

Montelle, Clemency and Kim Plofker (2013). "*Karaṇakesarī* of Bhāskara: A 17th-Century Table Text for Computing Eclipses". In: *History of Science in South Asia* 1, pp. 1–63.

Montelle, Clemency and Kim Plofker (2015). "The Transformation of a Handbook into Tables: The *Brahmatulyasāraṇī* and the *Karaṇakutūhala* of Bhāskara". In: *SCIAMVS* 16, pp. 1–34.

Montelle, Clemency and Kim Plofker (2018). *Sanskrit Astronomical Tables*. Cham: Springer.

Morgan, Daniel Patrick (2016). "The Planetary Visibility Tables in the Second-Century BC Manuscript Wu xing zhan 五星占". In: *East Asian Science, Technology, and Medicine* 43, pp. 17–60.

Nallino, Carlo Alfonso (1899–1907). *Al-Battani sive Albattenii Opus astronomicum* (*kitāb al-zīj al-sābi*) (vols. 1 and 2 reprinted by Minerva, Frankfurt 1969; vol. 3 reprinted by Olms, Hildesheim 1977). 3 vols. Milan: U. Hoeplium.

Neugebauer, Otto (1956). *Astronomical Cuneiform Texts* (2nd edition, New York: Springer, 1983). 3 vols. Princeton: Institute for Advanced Study.

Neugebauer, Otto and David Pingree (1970–1971). *The Pañcasiddhāntikā of Varāhamihira*. Copenhagen: Danish Royal Academy.

Neugebauer, Otto and David Pingree (1967). "The Astronomical Tables of Mahādeva". In: *Proceedings of the American Philosophical Society* 111, pp. 69–92.

Ossendrijver, Matthieu (2018). *Babylonian Mathematical Astronomy. Tabular Texts.* Sources and Studies in the History of Mathematics. New York: Springer.

Pandeya, Matri Prasada, ed. (1938). *Tithicintāmaṇi of Gaṇeśa Daivajña.* Haridas Sanskrit Series 76. Benares: Chowkhamba Sanskrit Series Office.

Pecchia, Cristina, ed. (forthcoming). *Editors of Sanskrit Texts.* Wien: Austrian Academy of Sciences Press.

Pedersen, Fritz Saaby (2002). *The Toledan Tables: A Review of the Manuscripts and the Textual Versions with an Edition.* Vol. 2. Historisk-filosofiske Skrifter 24. Four Parts. Copenhagen: Kgl. Danske Videnskabernes Selskab.

Pingree, David (1970–1994). *Census of the Exact Sciences in Sanskrit.* Series A. 5 vols. Philadelphia: American Philosophical Society.

Pingree, David (1968). "Sanskrit Astronomical Tables in the United States". In: *Transactions of the American Philosophical Society New Series* 58.3, pp. 1–77.

Pingree, David (1972). "Precession and Trepidation in Indian Astronomy before A.D. 1200". In: *Journal for the History of Astronomy* iii, pp. 27–35.

Pingree, David (1973). *Sanskrit Astronomical Tables in England.* Madras: Kuppuswami Sastri Research Institute.

Pingree, David (1978). "History of Mathematical Astronomy in India". In: *Dictionary of Scientific Biography.* Vol. 15. Detroit: Charles Scribner's Sons, pp. 533–633.

Pingree, David ed. (1987). *The Rājamṛgāṅka of Bhojarāja.* Aligarh Oriental Series 7. Aligarh: Viveka Publications.

Pingree, David (1996). "Bīja-Corrections in Indian Astronomy". In: *Journal for the History of Astronomy* 27, pp. 161–172.

Pingree, David (2003). *A Descriptive Catalogue of the Sanskrit Astronomical Manuscripts preserved at the Maharaja Man Singh II Museum in Jaipur, India.* Philadelphia: American Philosophical Society.

Pingree, David (2005). *Dorotheus of Sidon: Carmen Astrologicum.* Abingdon, MD: Astrology Classics Publishers.

Pingree, David (2007). *A Catalogue of the Sanskrit Manuscripts at Columbia University.* New York: American Institute of Buddhist Studies at Columbia University in the City of New York.

Plofker, Kim (2016). "*Siddhānta-karaṇa* Conversion: Some Algorithms in the *Grahaganitādhyāya* of Bhāskara II's *Siddhāntaśiromaṇi* and in his *Karaṇakutūhala*". In: *Gaṇita Bhāratī* 38, pp. 93–110.

Plofker, Kim (n.d.). "The 'Wonderful Astronomical Handbook' of Bhāskara and Ekanātha's Commentary". Forthcoming.

Plofker, Kim and Toke L. Knudsen (2011). "Calendars in India". In: *Calendars and Years: Astronomy and Time in the Ancient and Medieval World.* Ed. by John Steele. Oxford: Oxbow Books, pp. 53–68.

Rai, R.N. (1974). "The Prime Meridian in Indian Astronomy". In: *Indian Journal of History of Science* 9.1, pp. 45–50.

Rao, A.B. Padmanabha (2017). "Ideas of Infinitesimals of Bhāskarācārya in *Līlāvatī* and *Siddhāntaśiromaṇi*". In: *Indian Journal of History of Science* 52.1, pp. 17–27.

Rao, S. Balacandra and S.K. Uma (2008). *Karaṇakutūhalam of Bhāskarācārya II. An English Translation with Notes and Appendices*. New Delhi: Indian National Science Academy.

Robins, William (2007). "Editing and Evolution". In: *Literature Compass* 4.1, pp. 89–120.

Robinson, P. (2016). "Social Editions, Social Editing, Social Texts." In: *Digital Studies/Le champ numérique* 6. http://doi.org/10.16995/dscn.6.

Rochberg, Francesca (2016). *Before Nature: Cuneiform Knowledge and the History of Science*. Chicago: University of Chicago Press.

Rocher, Ludo (1995). "Sanskrit Literature". In: *Scholarly Editing: A Guide to Research*. Ed. by D.C. Greetham. New York: Modern Language Association of America, pp. 575–599.

Ruggles, Clive L.N., ed. (2015). *Handbook of Archaeoastronomy and Ethnoastronomy*. 3 vols. New York: Springer.

Rupa K., Padmaja Venugopal, and S. Balacandra Rao (2014). "*Makaranda Sāriṇī* and Allied *Saurapaksa* Tables—A Study". In: *Indian Journal of History of Science* 49.2, pp. 186–208.

Saha, Shandip (2004). "Creating a Community of Grace: A History of the Puṣṭi Mārga in Northern and Western India (1493–1905)". PhD thesis. University of Ottawa.

Salomon, Richard (1998). *Indian Epigraphy. A Guide to the Study of Inscriptions in Sanskrit, Prakrit, and the Other Indo-Aryan Languages*. New York/Oxford: Oxford University Press.

Samsó, Julio (2003). "Review of Fritz D. Pedersen: *The Toledan Tables. A Review of the Manuscripts and the Textual Versions with an Edition*". In: *Suhayl* 3, pp. 246–251.

Sarma, K.V. (2013). "Makaranda". In: *Encyclopaedia of the History of Science, Technology, and Medicine in Non-Western Cultures*. Ed. by Helaine Selin. New York: Springer, pp. 547–548.

Sarma, Sreeramula Rajeswara (2019). "The Legend of *Līlāvatī*". In: *Bhāskara-prabhā*. Ed. by Takao Hayashi, Clemency Montelle, and K. Ramasubramanian. Culture and History of Mathematics 11. Hindustan Book Agency, pp. 23–39.

Śāstrī, Bāpū Deva, ed. (1989). *The Siddhāntaśiromaṇi, a Treatise on Astronomy*. 2nd edition. Kashi Sanskrit Series 72. Varanasi: Chaukhambha Sanskrit Sansthan.

Sewell, Robert and S.B. Dikshit (1896). *The Indian Calendar*. London: Swan Sonnenschein.

Sircar, D.C. (1965). *Indian Epigraphy* (reprinted 1996). Delhi: Motilal Banarsidass.

Slaje, Walter, ed. (2008). *Śāstrārambha: Inquiries into the Preamble in Sanskrit*. Abhandlungen für die Kunde des Morgenlandes 62. Germany: Harrassowitz Verlag.

Smith, Frederick M. (2011). "Predestination and Hierarchy: Vallabhācārya's Discourse on the Distinctions between Blessed, Rule-bound, Worldly, and Wayward Souls (the *Puṣṭipravāhamaryādābheda*)". In: *Journal of Indian Philosophy* 39, pp. 173–227.

Tables Analysis Method for the history of Astral Sciences (2015). https://tamas.hypotheses.org/.

Tanselle, G.T. (1996). "The Varieties of Scholarly Editing". In: *Scholarly Editing: A Guide to Research*. Ed. by D.C. Greetham. New York: Modern Language Association of America, pp. 9–32.

Tihon, Anne (2011). *Les Tables Faciles de Ptolémée: 1a Tables A1–A2. Introduction, édition critique*. Louvain-La-Neuve: Université Catholique de Louvain/Peeters.

Trovato, Paolo (2014). *Everything You Always Wanted to Know about Lachmann's Method*. Padua: libreriauniversitaria.it edizioni.

Van Brummelen, Glen (2014). "The Travels of Astronomical Tables within Medieval Islam: A Summary". In: *Suhayl* 13, pp. 11–21.

van Dalen, Benno (1993). "Ancient and Medieval Astronomical Tables. Mathematical Structure and Parameter Values". PhD thesis. Netherlands: Utrecht University.

Van Hulle, Dirk (n.d.). *Annotated Bibliography: Key Works in the Theory of Textual Editing*. https://www.mla.org/Resources/Research/Surveys-Reports-and-Other-Documents/Publishing-and-Scholarship/Reports-from-the-MLA-Committee-on-Scholarly-Editions/Annotated-Bibliography-Key-Works-in-the-Theory-of-Textual-Editing.

West, Martin Litchfield (1973). *Textual Criticism and Editorial Technique Applicable to Greek and Latin Texts*. Bibliotheca scriptorum Graecorum et Romanorum Teubneriana. Stuttgart: Teubner.

Worthington, Martin (2012). *Principles of Akkadian Textual Criticism*. Berlin: De Gruyter.

Wujastyk, Dominik (1993). *Metarules of Pāṇinian Grammar: Vyāḍi's Paribhāṣāvṛtti*. 2 vols. Groningen: Egbert Forsten.

Yano, Michio (1997). "Distance of Planets in Indian Astronomy". In: *Oriental Astronomy from Guo Shoujing to King Sejong*. Ed. by Is-Seong Nha and F. Richard Stephenson. Seoul: Yonsei University Press, pp. 113–120.

Yano, Michio (2003). "Calendar, Astrology, and Astronomy". In: *Blackwell Companion to Hinduism*. Ed. by Gavin Flood. Oxford: Blackwell, pp. 376–392.

Yano, Michio (2004). *Pancanga (Version 3.14)*. http://www.cc.kyoto-su.ac.jp/~yanom/pancanga/index.html.

Index of Names and Subjects

www.ingramcontent.com/pod-product-compliance
Lightning Source LLC
LaVergne TN
LVHW020635300926
849161LV00004B/719